NUREG-1710
Vol. 2

History of Water Development at the Nevada Test Site: A Literature Review

I0494111

U.S. Nuclear Regulatory Commission
Advisory Committee on Nuclear Waste
Washington, DC 20555-0001

AVAILABILITY OF REFERENCE MATERIALS
IN NRC PUBLICATIONS

NRC Reference Material

As of November 1999, you may electronically access NUREG-series publications and other NRC records at NRC's Public Electronic Reading Room at http://www.nrc.gov/reading-rm.html.
Publicly released records include, to name a few, NUREG-series publications; *Federal Register* notices; applicant, licensee, and vendor documents and correspondence; NRC correspondence and internal memoranda; bulletins and information notices; inspection and investigative reports; licensee event reports; and Commission papers and their attachments.

NRC publications in the NUREG series, NRC regulations, and *Title 10, Energy*, in the Code of *Federal Regulations* may also be purchased from one of these two sources.
1. The Superintendent of Documents
    U.S. Government Printing Office
    Mail Stop SSOP
    Washington, DC 20402–0001
    Internet: bookstore.gpo.gov
    Telephone: 202-512-1800
    Fax: 202-512-2250
2. The National Technical Information Service
    Springfield, VA 22161–0002
    www.ntis.gov
    1–800–553–6847 or, locally, 703–605–6000

A single copy of each NRC draft report for comment is available free, to the extent of supply, upon written request as follows:
Address: Office of the Chief Information Officer,
           Reproduction and Distribution
             Services Section
           U.S. Nuclear Regulatory Commission
           Washington, DC 20555-0001
E-mail: DISTRIBUTION@nrc.gov
Facsimile: 301–415–2289

Some publications in the NUREG series that are posted at NRC's Web site address http://www.nrc.gov/reading-rm/doc-collections/nuregs are updated periodically and may differ from the last printed version. Although references to material found on a Web site bear the date the material was accessed, the material available on the date cited may subsequently be removed from the site.

Non-NRC Reference Material

Documents available from public and special technical libraries include all open literature items, such as books, journal articles, and transactions, *Federal Register* notices, Federal and State legislation, and congressional reports. Such documents as theses, dissertations, foreign reports and translations, and non-NRC conference proceedings may be purchased from their sponsoring organization.

Copies of industry codes and standards used in a substantive manner in the NRC regulatory process are maintained at—
        The NRC Technical Library
        Two White Flint North
        11545 Rockville Pike
        Rockville, MD 20852–2738

These standards are available in the library for reference use by the public. Codes and standards are usually copyrighted and may be purchased from the originating organization or, if they are American National Standards, from—
        American National Standards Institute
        11 West 42nd Street
        New York, NY 10036–8002
        www.ansi.org
        212–642–4900

Legally binding regulatory requirements are stated only in laws; NRC regulations; licenses, including technical specifications; or orders, not in NUREG-series publications. The views expressed in contractor-prepared publications in this series are not necessarily those of the NRC.

The NUREG series comprises (1) technical and administrative reports and books prepared by the staff (NUREG–XXXX) or agency contractors (NUREG/CR–XXXX), (2) proceedings of conferences (NUREG/CP–XXXX), (3) reports resulting from international agreements (NUREG/IA–XXXX), (4) brochures (NUREG/BR–XXXX), and (5) compilations of legal decisions and orders of the Commission and Atomic and Safety Licensing Boards and of Directors' decisions under Section 2.206 of NRC's regulations (NUREG–0750).

NUREG-1710
Vol. 2

History of Water Development at the Nevada Test Site: A Literature Review

Manuscript Completed: May 2004
Date Published: February 2005

Prepared by
M.P. Lee, N.M. Coleman

U.S. Nuclear Regulatory Commission
Advisory Committee on Nuclear Waste Staff
Washington, DC 20555-0001

NUREG-1710, Volume 2, has been
reproduced from the best available copy.

ABSTRACT

Historic accounts, geologic treatises, and other key literature sources were used to chronicle developments in the Nevada Test Site (NTS) during the past 150 years. As was the case in the nearby Amargosa Desert, human activities in the area currently occupied by NTS were initially influenced by the location of cold springs. They provided indigenous Native Americans with drinking water. Later, as part of the Western expansion, many of these same springs were relied on by Euro-American pioneers as they crossed the continent. By the time NTS was engaged in activities related to the Nation's defense, it was necessary to develop the available subsurface ground-water supplies, aided in part by improved geologic knowledge of local resources. The first well supporting this infrastructure was Army Well No. 1. The 1253-foot well was completed in May 1956. Today, 17 wells distributed among four service areas supply NTS water needs. Most were drilled in the mid-to-late 1950s or early 1960s in Yucca Flat, Frenchman Flat, and Mercury Valley. Overall, the welded volcanic tuff aquifer is only locally important (in Jackass Flats) whereas the lower carbonate aquifer serves other portions of NTS.

This report is the second volume in the NUREG-1710 series.

CONTENTS

Abstract iii

Conversion Factors vii

Acknowledgments ix

Abbreviations xi

1. Introduction 1
    1.1 Purpose of Report 1
    1.2 Geographic Setting 2
    1.3 Geologic Setting 2
    1.4 Hydrology 4
         1.4.1 Cold Springs 5
         1.4.2 Tanks and Playas 5

2. Historical Developments 9
    2.1 Native American Activities 9
    2.2 Homesteading 12
    2.3 Mining 14
    2.4 Land Withdrawals: 1930s–1950s 17
    2.5 Atomic Programs 17

3. References 25

FIGURES

1. Current Boundaries and Subdivisions within NTS 3

2. Hydrographic Areas within or adjacent to NTS 6

3. Springs, Tanks, and Playas within NTS 7

4. Map Showing Location of Recorded Cultural Resources within NTS 11

5. Wells within NTS 20

6. Completion Details for Well J-12 23

7. Completion Details for Well J-13 24

TABLES

1. Types of Cultural Sites Found within NTS 10

2. Possible Relay Stations within NTS 13

3. Reported Mining Districts within NTS 16

4. Possible Ranching Locations within NTS 18

5. Perennial Yields and Peak Historic Water Demand for NTS Hydrographic Areas 22

APPENDICES

A. Major Nevada Test Site Facilities and Water-Supply Source Areas A-1

B. Summary of Well Information for Nevada Test Site Service Areas B-1

CONVERSION FACTORS

The preferred system of measurement today is the metric system, or the "Systèm Internationale (SI)." However, for some physical quantities, many scientists and engineers (as well as drillers) prefer the familiar, and continue to use inch/pound units (the so-called U.S. customary system). Therefore, for ease of comparison with existing drilling practice, in this regard, inch/pound units will be used in this report.

Multiply	By	To obtain
inch (in)	2.54	centimeter
feet (ft)	0.3048	meter
mile (mi)	1.609	kilometer
square mile (mi^2)	2.590	square kilometer
gallon	0.00379	cubic meter
gallons per day (gpd)	0.00379	cubic meters per day
gallons per minute (gpm)	0.00379	cubic meters per minute

Temperature in degrees Fahrenheit (°F) may be converted to degrees Celsius (°C) as follows:

$$°C = (°F - 32)/1.8.$$

ACKNOWLEDGMENTS

The authors wish to thank several individuals who provided useful information, comments, and reviews that have enhanced the substance of this report as it was being developed. These individuals include:

David Brooks (NMSS)
Jeff Ciocco (NMSS)
Bret Leslie (NMSS)
Larry McKague (CNWRA)
Tom Nicholson (RES)
Gordon Wittmeyer (CNWRA)

In addition, several individuals were helpful in locating some of the references cited in this report. They include Harold Drollinger (Desert Research Institute/Las Vegas) and Chad Glenn (NMSS). Ellen Kraus (NMSS) provided editorial guidance.

These individuals do not necessarily approve, disapprove, or endorse the views expressed in this report. The views expressed herein are the authors' and do not reflect an NRC staff position, or any judgment or determination by the Advisory Committee on Nuclear Waste or the NRC, regarding the matters addressed or the acceptability of a license application for a geologic repository at Yucca Mountain.

ABBREVIATIONS

ACNW   Advisory Committee on Nuclear Waste
AEC   Atomic Energy Commission

CNWRA   Center for Nuclear Waste Regulatory Analyses
CRWMS M&O   Civilian Radioactive Waste Management System
                Management and Operating (contractor to DOE)

DOE   U.S. Department of Energy

EPA   U.S. Environmental Protection Agency
ERDA   Energy Research & Development Administration

HLW   high-level radioactive waste (including spent nuclear fuel)

NAFR   Nellis Air Force Range Complex
NBMG   Nevada Bureau of Mines and Geology
NMSS   Office of Nuclear Material Safety and Safeguards (within NRC)
NRC   U.S. Nuclear Regulatory Commission
NRDS   Nuclear Rocket Development Station
NTS   Nevada Test Site

RES   Office of Nuclear Regulatory Research (within NRC)

T and R   township and range

URL   uniform resource locator (on the Internet)
USGS   U.S. Geological Survey

1. INTRODUCTION

Water availability has been a long-standing issue in the history of the arid Southwest United States (Reisner, 1986). Natural precipitation is generally insufficient for farming. Most water supplies in Western Nevada originate in the Sierra Nevada Mountains with little, if any, generated in the lowlands. There are few perennial or intermittent streams present. For example, the only major occurrences of surface water, in Southern Nevada, are cold springs and the *Colorado River* (and its impoundments). There are many ephemeral stream channels present, such as the *Amargosa River*. These channels include those associated with the major drainage systems. Springs cause short reaches of the Amargosa River (i.e., the *Carson Slough*) to be perennial to intermittent in their flow. Runoff occurs irregularly in response to both convective summer and frontal winter storms, with stream discharge rates varying greatly in magnitude.

On arriving in the arid Southwest, Euro-American pioneers initially adopted Native American practices to supply fresh water needs. Later, during the Industrial Revolution, water-management practices from the mining industry were modified to supply the agricultural and municipal water requirements of growing western communities. Advances in drilling and pumping technology were adopted from the petroleum and natural gas industry at the beginning of this century. These advances, combined with increased geologic knowledge, were ultimately instrumental in exploiting heretofore inaccessible subsurface, ground-water resources. The history of water development in the Amargosa Desert area generally follows the model typical of the Southwestern United States.

1.1 Purpose of Report

The purpose of this report is to summarize the history of water development within the Nevada Test Site (NTS). In light of the potential geologic repository for the disposal of spent nuclear fuel and other high-level radioactive waste (HLW) at the Yucca Mountain site, understanding those water-development factors that have influenced the pattern and growth in the area has been studied. Although factors affecting water development have been examined in the past for several arid Western states (e.g., Baker and others, 1973), based on a review of the literature, such factors have not been examined for Nevada, in general, and NTS in particular. Consequently, the authors, while formerly members of the U.S. Nuclear Regulatory Commission (NRC) staff, performed a review of the literature, both printed and electronic, to identify key references. After completion of that review, it was determined that it would be useful to summarize this information in a synthesis report, as a way to preserve institutional knowledge on this subject.

Also, it should be noted the history of the area has been the subject of previous study, for other reasons [e.g., Steward (1938); Worman (1969); Pippin and Zerga (1983); Lingenfelter (1986); McCracken (1990 and 1992); Stoffle and others (1990); Myrick (1992); Drollinger and others (1999); and Hartwell and Valentine (2002)]. The Energy Research & Development Administration (ERDA – 1977) and the U.S. Department of Energy (DOE) conducted environmental assessments over the years (DOE, 1984, 1986, 1996, 2002), for which some local information on water use was reported. Land and water-use practices were the subject of preliminary study by DOE (1988) following the designation of Yucca Mountain by Congress as a candidate site for a mined geologic repository. More recently, the U.S. Environmental Protection Agency (EPA-1999) continued to examine land and water-use practices as part of the development of its radiation protection standards for the proposed geologic repository. In preparing this report, the authors relied on these and other key sources to construct a chronological framework for crafting the discussions. No proprietary or unpublished data sources were reviewed — only information in the public domain has been cited.

This report is the second volume in the series. In Volume 1 of NUREG-1710 (Lee and others, 2005), historic accounts, geologic treatises, and other literature sources were used to chronicle water-use developments in the Amargosa Desert area during the past 150 years. In this installment, designated Volume 2, similar developments affecting water use within NTS are described.

Finally, the views expressed herein are the authors'. They do not reflect an NRC staff position, or any judgment or determination by the Advisory Committee on Nuclear Waste or the NRC, regarding the matters addressed or the acceptability of a license application for a geologic repository at Yucca Mountain.

The use of firm, trade, and brand names in this NUREG is for identification purposes and does not constitute endorsement by the NRC.

1.2   Geographic Setting

NTS is located 65 mi northwest of Las Vegas, in Nye County. Geographically, it may be considered the most dominant feature in Nevada, covering approximately 1375 mi^2 – a land area larger than the State of Rhode Island. This remote site is surrounded by thousands of additional acres of land withdrawn from the public domain for use as a protected wildlife range and for a military gunnery range (the Nellis Air Force Range Complex — NAFR), thereby creating an unpopulated area comprising some 5470 mi^2 (Figure 1). Public access to both NTS and NAFR is limited.

Initially established as the Atomic Energy Commission's (AEC's) on-continent proving ground, NTS has seen more than four decades of nuclear weapons testing. Since the nuclear weapons testing moratorium in 1992 and under the direction of DOE, the test site use has diversified into many other government programs within 27 designated "areas." Although not publicly accessible, NTS has about 400 mi of paved roads, 300 mi of unpaved roads, two airstrips, and 10 heliports, as well as electric power transmission and water supply systems. The site has more than 1100 permanent buildings, including a hospital, post office, fire station, sheriff's substation, and movie theater; a motor pool (complete with a repair complex); training facilities, and waste water treatment plant. Overall, the estimated value of capital improvements to the NTS site is about $700 million. In the late 1960s, NTS employment was as high as 8000 personnel (Miller, 1968; p. 2). In the 1990s, employment at NTS was estimated to be about 3000 personnel (TRW Environmental Safety Systems, Inc., 1995; p. 4-3).

Finally as regards climate, Nevada is the most arid State in the conterminous United States (Geraghty and others, 1973; Plate 3) – the present climate is classified as a mid-latitude desert. Local temperatures range at NTS from an average daily minimum of about 27°F, in January, to an average daily maximum of over 99°F in July, with wide daily and seasonal variations. The annual precipitation is less than 6 in. The distribution of precipitation is related to the elevation of the land surface and the latitude. Precipitation is usually lower in the valleys and higher in the surrounding mountain ranges and mesas. Most of the precipitation falls in the winter, although summer thunderstorms are not uncommon. The humidity is low and wind movement moderate most of the year.

1.3   Geologic Setting

NTS is located in the Basin and Range physiographic province (Dohrenwend, 1987). This region is distinguished by its topographically-closed (intermontane) basins, which are separated by mountain ranges,

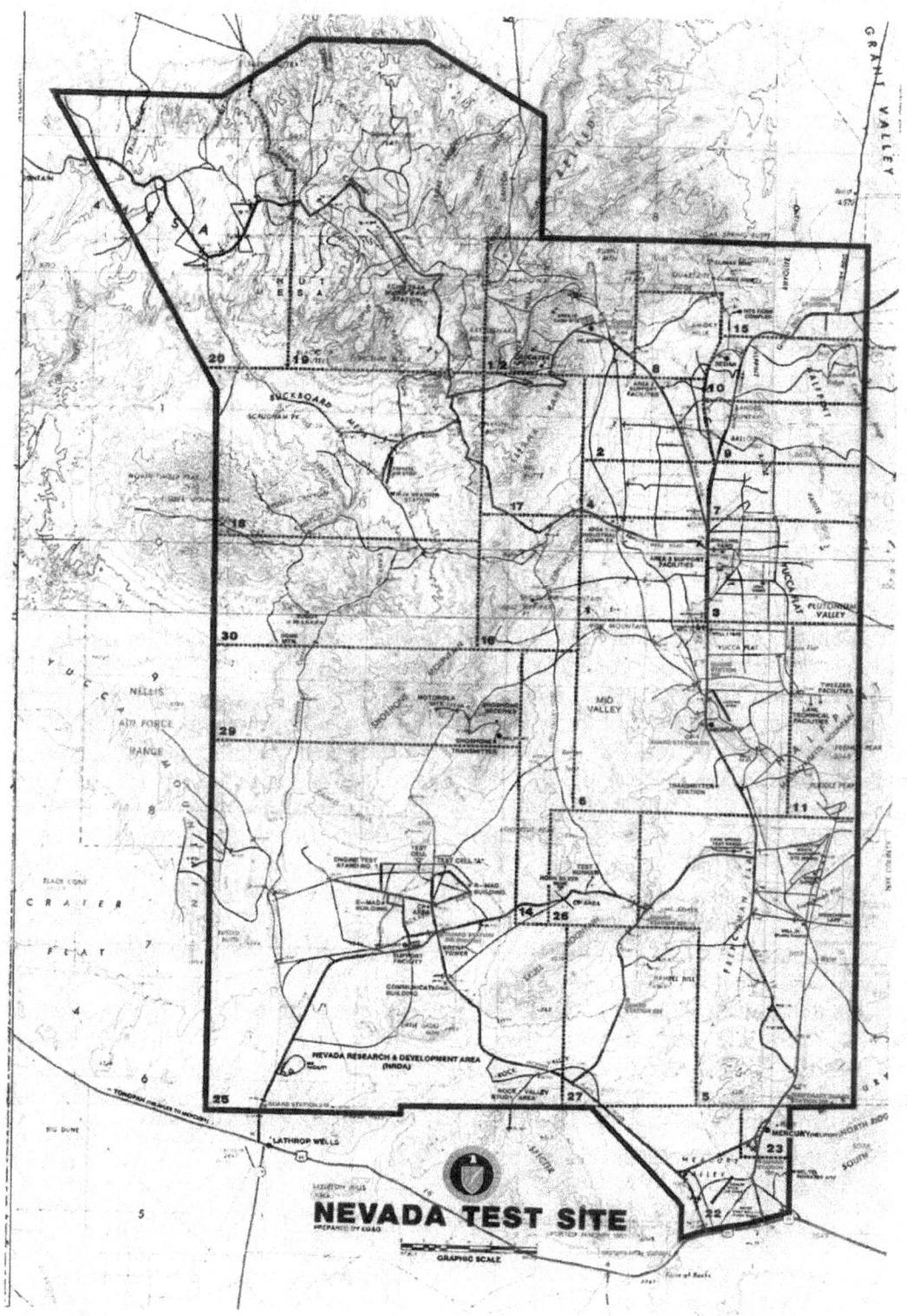

Figure 1. **Current Boundaries and Subdivisions within NTS.** The NTS subdivisions are referred to as "Areas." Figure available from DOE.

3

hills, and mesas, with internal drainage into the basins (valleys). The mountain ranges average from 5 to 15 mi in length but can exceed 50 mi, with elevations of 1000 to 11,000 ft above the valley floors. By contrast, altitudes of the valley floors range from below sea level (*Death Valley*) to about 5500 ft (*Kawich Valley*). Within NTS, elevations range from 3080 ft at *Jackass Flats* (a basin), in the southwest corner of the site, to 7675 ft at *Rainier Mesa*, toward the northern boundary of NTS.

Rocks in the northern and southwestern portions of NTS are principally volcanics: basalts, rhyolites, and tuffs. The mountains are the principal areas of erosion and are characterized by relatively barren, steep topography. Elsewhere, the basins are filled with Tertiary- and Quaternary-age alluvial deposits. They range from semi- to unconsolidated, including the following major lithologies: river-channel and playa lake deposits, alluvial fan sediments; eolian sands, and Tertiary-age conglomerates. Sometimes sequences of ash-fall tuffs are present. Overall, the alluvial-fill deposits range in thickness from an average of 2000 ft to about 4500 ft.

Mountainous areas generally have a thin soil veneer because of weathering in-situ and have a low moisture-holding capacity. Soils within the alluvial basins were formed on recent alluvium deposited by the Amargosa River and its tributaries. The soils found in the alluvial basin are medium- to fine-textured, with somewhat higher-moisture holding capacities. Because they have a low organic matter content and drain rapidly, these soils are generally considered nonarable and not suitable for farming.

1.4 Hydrology [1]

The principal topographic features in the southern end of NTS are the ephemeral stream channels of the Amargosa River and its tributaries. This river basin system generally defines the geographic extent of the *Amargosa Desert*.[2] The land area also approximately corresponds to the *Amargosa Desert hydrographic area* used to define hydrographic regions, areas, or basins subject to water-resource investigations in Nevada (Rush, 1968). This basin covers an area of about 900 mi^2 in Nye County and 470 mi^2 in Inyo County, and is the largest of nine subbasins of the greater *Death Valley hydrographic area* (Figure 2).

Short reaches of the Amargosa River near Beatty and south of Amargosa Farms to Shoshone (California) are perennial to intermittent in flow because of local cold springs or the intersection of the water table with the ground surface (French and others, 1984; Waddell and others, 1984). Some places contain wet playas from which ground water is continuously discharged by evapo-transpiration. The Amargosa River and its tributaries, therefore, only carry significant amounts of water after periods of infrequent but intense precipitation – in the form of cloudbursts or thunderstorms.

There are no perennial streams within NTS. *Fortymile Wash* is the principal drainage feature within NTS. But it, like most of the other tributaries within the Amargosa River system, is ephemeral. Thus, before development of ground-water resources, the only reliable water supplies within NTS were

---

[1]This section is intended to provide the reader with additional background on water availability in the NTS area. It is not designed to be a comprehensive review of the geologic literature. Instead, the goal here is to identify what physical factors influence the occurrence and availability of fresh water, and in doing so provide the reader with an appreciation of the timing and the manner of how this resource within the site was developed.

[2]Sometimes more commonly referred to as the *Amargosa Valley*.

the many cold springs, as well as a few natural tanks.

1.4.1 Cold Springs

Cold springs are located sporadically but frequently in and around NTS.[3] Thordarson and Robinson (1971) reported that there are about 750 springs within 100 mi of the site. Springs occur when the water table or a perched ground-water body intersects the ground surface. Changes in structure, lithology, and/or topography lead to changes in aquifer permeability, causing spring localization. Springs are typically of two types (Bryan, 1919): the *contact variety* – where a permeable rock overlies a rock of much lower permeability, such as a contact between the more permeable alluvial deposits and less permeable bedrock, or the *depression variety* – in which ground water seeps into topographic depressions that are covered with a veneer of detrital material, usually gravel.

Most springs display variations in rates of flow or discharge. Flow rates can be constant or variable (seasonal). Most springs in the area discharge around 10 gpm or less (Thordarson and Robinson, 1971). So-called perennial springs (those associated with the water table) have generally constant flow rates, whereas ephemeral or "wet-weather" springs are likely to have more variable flow rates because of their close proximity to recharge areas. In the Yucca Mountain area, for example, perched water occurs primarily in the foothills and ridges that flank the basins (French and others, 1981; p. 49) and creates flow in ephemeral springs in direct response rainfall or snowmelt.[4]

Springs were the earliest sources of water supply to both the indigenous inhabitants of the area as well as the more recent Euro-American arrivals. Their location in arid areas can be easily identified by the general abundance of unique types of foliage and wildlife (Ball, 1907; pp. 22–24) —so-called *riparian zones* —when compared with their surroundings.[5] In general, standing or pooled water can be found throughout the year at several of the higher-discharging spring locations.

Major spring locations within portions of NTS are shown in Figure 3.

1.4.2 Tanks and Playas

Tank are naturally occurring cisterns found in impervious rock. Most can be found at higher elevations, within the boundaries of NTS. They are impermeable, topographic depressions that form natural collection basins for precipitation and snow. The volume and quality of water in tanks typically decrease in the summer months. By August, many tanks within NTS are dry and the water in others is scarcely drinkable (Ball, 1907; pp. 23–24). Major tank locations within NTS are also shown in Figure 3.

Playas also occur in the *Frenchman Flat*, *Franklin Lake*, *Gold Flat*, *Kawich Valley*, and *Yucca Flat* areas, and they collect surface-water run-off from periodic thunderstorms. Because thunderstorms are infrequent and because the surface water typically evaporates within a few days or weeks, playas have never been considered a reliable source of water supply.

---

[3]Also includes *weeps* and *seeps* – springs with low discharge rates, typically too small to measure, and *phreatophyte areas* – locations of plants with deep root systems capable of obtaining their water from the saturated zone. The principal *phreatophyte areas* can be found in the Amargosa Valley (Walker and Eakin, 1963; p. 22).

[4]The aquifers supplying ephemeral springs are perched aquifers, separated from the water table by a layer of reduced permeability (aquitard). Because the amount of water they discharge fluctuates during periods of high

precipitation, it has been suggested that many of the perched springs possess juvenile, meteoric water (Ball, 1907; p. 21).

[5]Native Americans or later explorers often erected stone markers to mark their locations (*Op cit.*).

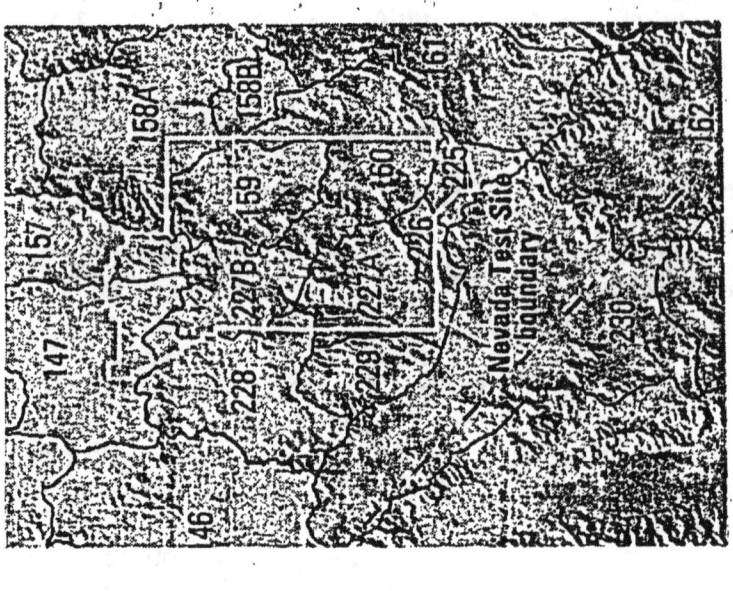

NDWR Hydrographic Area	Area Number on Map	Approximate Area (sq mi)	Approximate Altitude (ft)
Gold Flat	147	684	5200
Kawich Valley	157	350	5500
Emigrant Valley	158	767	4600
Yucca Flat	159	305	4000
Frenchman Flat	160	463	3200
Mercury Valley	225	225	3200
Rock Valley	226	82	3300
Fortymile Canyon; Jackass Flats	227A	279	3500
Buckboard Mesa	227B	240	5000
Oasis Valley	228	460	3800
Crater Flat	229	182	3200
Amargosa Desert	230	890	2600

Figure 2. Hydrographic Areas within or adjacent to NTS. Hydrographic area boundaries shown in the figure are defined by the Nevada Division of Water Resources (NDWR). Summary table data taken from Rush (1968). Figure (scale 1" ≈ 42 mi) taken from Hevesi and others (2003, p. 9). See the USGS URL site at *http://water.usgs.gov/pubs/wri/wri03409ll*.

6

SPRINGS (S) ●	
S1	Tub
S2	Oak
S3	Ranier Mesa
S4	White Rock
S5	Captain Jack
S6	Tippipah
S7	Reitman Seep
S8	Unnamed
S9	Twin
S10	Topopah
S11	Yellow Rock
S12	Cane
S13	Pavits
S14	Tupapa Seep
TANKS (T) ▲	
T1	Unnamed
T2	Small
T3	Ammonia
T4	Triple (Railroad)
T5	Iron
PLAYAS (P) ■	
P1	Yucca (Tippipah)
P2	Frenchman

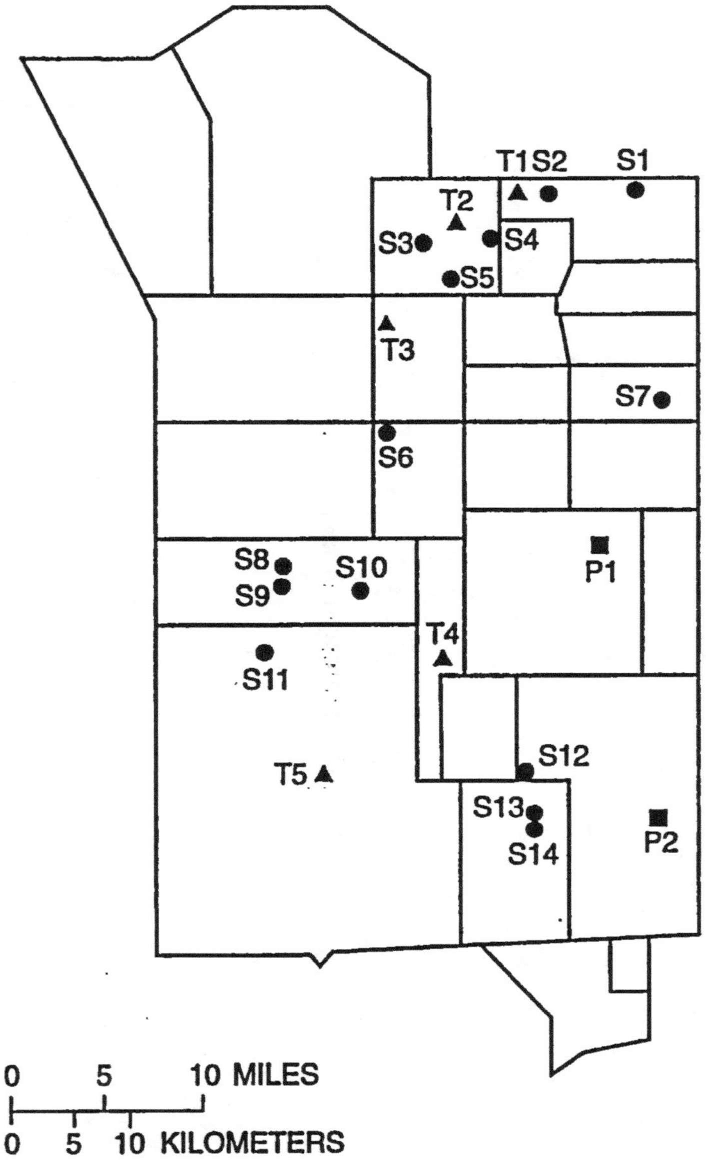

Figure 3. Springs, Tanks, and Playas within NTS. Some of the springs and tanks identified are described in Ball (1907) and Thordarson and Robinson (1971, Table 4).

7

This page is intentionally left blank.

2. HISTORICAL DEVELOPMENTS

As was the case in the nearby Amargosa Desert, the literature suggests that human activities in the area currently occupied by NTS were initially influenced by the location of cold springs. They provided indigenous Native Americans with drinking water as they foraged for food. Later, as part of the Western expansion, many of these same springs were relied on by Euro-American pioneers as dependable sources of fresh water as they crossed the continent. During the first half of this century, prospectors and newer homesteaders continued to rely on these springs. However, by the time NTS was engaged in activities related to the Nation's defense, the discharge from these springs proved inadequate and the subsurface ground-water resource beneath the site was developed.

2.1 Native American Activities

In historic times, Southern Nevada (including the area containing the Amargosa Desert and NTS) was occupied by the Western Shoshone and Southern Paiute.[6] These peoples were descendants of Paleo-Native American cultures believed to have occupied the region during the preceding 10,000 years. The Western Shoshone and Southern Paiute were regarded as nomadic hunter-foragers who relied on the numerous springs found throughout the area as well as for the animals and vegetation these springs attracted and sustained (Stoffle and others, 1990). Ball (1907, pp. 22-23) notes that the placement of many of the early Indian trails in the area was influenced by the locations of the various springs and tanks, and the distance between these sites rarely exceeded 40 mi (Op cit.). Thordarson and Robinson (1971, Table 4) identified about 80 springs of various sizes in Southern Nye County, in and around NTS (south of the 37°15' parallel).

The literature indicates that Native Americans lived within the current NTS site and regularly traversed it while gathering piñon nuts (a staple) and hunting for game. Investigations have shown evidence of the Southern Paiute culture in caches of artifacts (beads, pottery, etc.) at camp sites, rock shelters, stone circles, and other archeological sites within NTS. Over 1700 cultural resource sites have been identified and cataloged within NTS (Table 1). Most of these sites are generally located to the east and to the north of the proposed repository site – in the vicinity of *Fortymile Canyon* and *Buckboard* and *Pahute Mesas*, respectively (Figure 4). Many are in close proximity to major springs and tanks (Figure 3). Near Yucca Mountain, for example, the terraces adjacent to Fortymile Canyon contain abundant artifacts in the form of projectile points, blanks, and flakes (Worman, 1969; Pippin and Zerga, 1983). Several of the NTS spring locations had rock shelters, including food caches; but some of these locations have been interpreted by the literature for use principally during seasonal harvesting/hunting excursions (Worman, 1969; Stoffle and others, 1990). Manly (1927, p. 151) notes that California-bound miners (the "Forty-niners") encountered Southern Paiute growing maize and squash at a permanent camp at *Cane Springs*. Steward (1938, pp. 93-99, 182-185) also reports that at least nine Shoshone family (or family groups), totaling about 40 individuals, where in residence in a series of permanent camps in the vicinity of *White Rock Spring* as late as 1880. In general, though, the literature indicates that the function and role of many of the NTS archeological sites changed with time in response to the Western migration of Euroamericans.

By the 1900s, the size of the indigenous

---

[6]The reader is also referred to Section 2 ("Historical Developments Before 1900") of Volume 1 of this NUREG series for additional historical background on Native American activities in Southern Nevada, including lands now part of NTS.

Table 1. Types of Cultural Sites Found within NTS. The following reports the results of 501 archeological investigations to identify possible cultural resources within NTS. These investigations covered 40,383 acres or 5 percent of NTS and are indexed to NDWR hydrographic areas. The number and kinds of Native American sites can be used to evaluate the relative importance of a particular foraging area. *Residential Base Camps* have been interpreted to be locations of extended occupation, such as *Cane and Captain Jack Springs. Temporary camps* likely supported activities at the other localities listed. Taken from DOE (1996, pp. 4-155 – 4-159).

NDWR Hydrographic Area	Site Type								Totals
	Residential Base	Temporary Camp	Extractive Locality[a]	Processing Locality[b]	Unspecified Locality[c]	Cache[d]	Station[e]	Untyped	
Mercury Valley	0	0	0	0	3	0	0	0	3
Rock Valley	0	1	1	0	15	0	0	0	17
Fortymile Canyon/Jackass Flats	1	35	15	59	236	7	1	9	363
Buckboard Mesa	0	103	6	94	203	5	1	54	466
Oasis Valley	0	14	1	20	82	0	0	2	119
Gold Flat	0	25	1	96	124	10	0	1	257
Kawick Valley	0	9	0	25	37	0	0	8	79
Emigrant Valley	0	0	0	0	5	0	0	0	5
Yucca Flat	4	54	10	34	126	56	0	13	297
Frenchman Flat	1	2	2	38	52	0	0	0	95
TOTALS	6	243	36	366	883	78	2	87	1701

a. Resource procurement areas, such as quarries, water catchment basins, hunting blinds, or plant resource extraction locations.
b. Areas where resources, such as stone tools, plants, and animals were processed.
c. Lack of specific information to permit interpretation, as compared to *Residential Base Camps* or *Temporary Camps.*
d Temporary locations for storing resources or artifacts.
e. Special-purpose locations, including sites designated for the communication of information and rituals, including rock cairns, geoglyphs, observation points, and overlooks.

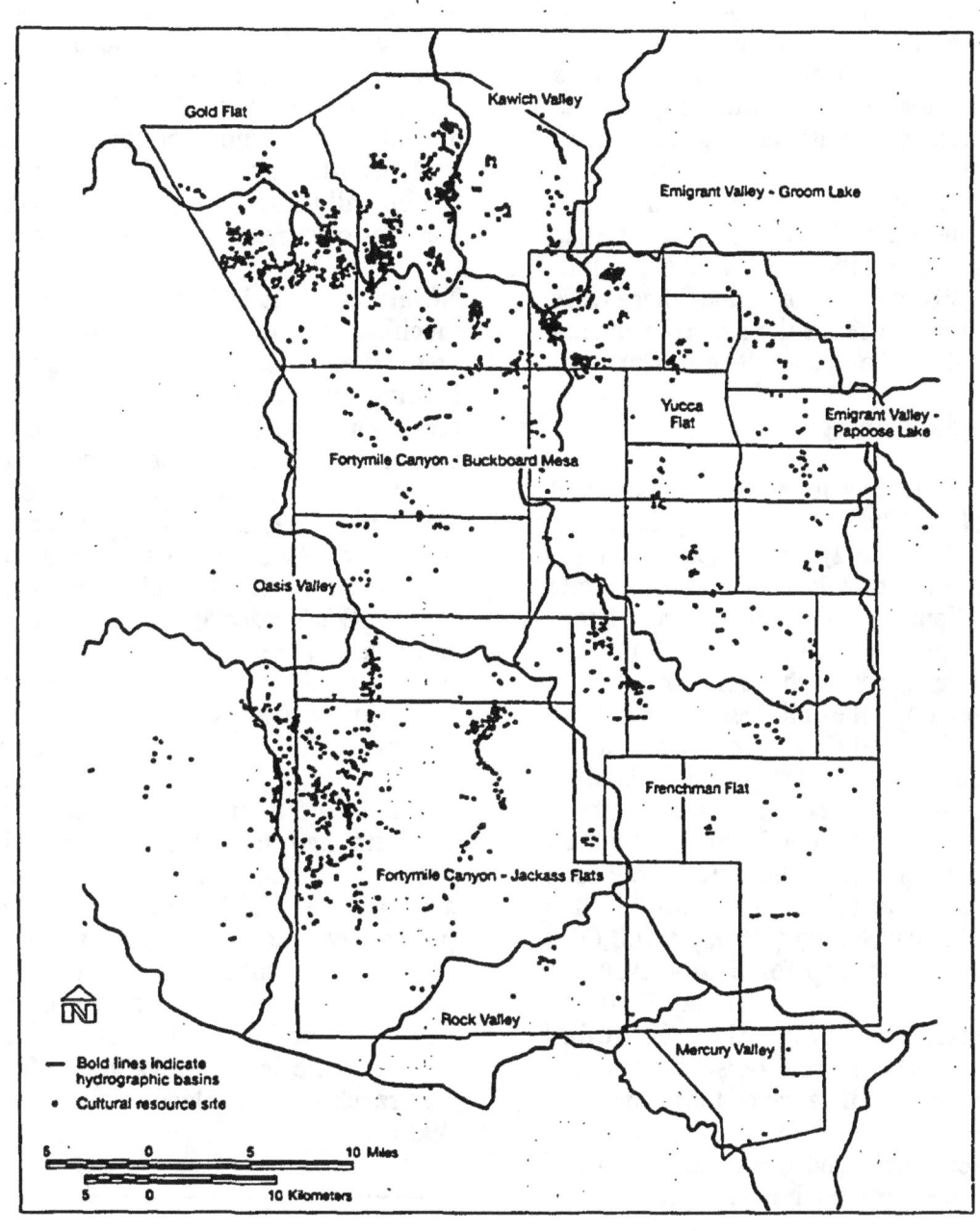

Figure 4. Map Showing Location of Recorded Cultural Resources within NTS. Taken from DOE (1996, Figure 4-47).

11

population had decreased for a variety of reasons. What had once been prime foraging and hunting areas used seasonally, because of its remoteness and lack of commercial potential, NTS lands had now become an unofficial Native American refuge (Stoffle and others, 1990; p. 114). More regular use would be consistent with the large number of sites interpreted in the literature as *Temporary Camps*. Native Americans occupied and maintained both a subsistence and residential relationship with NTS until the late 1930s, when public lands there were withdrawn from use for the first time during the outbreak of the Second World War.

2.2 Homesteading

Before 1900, relay stations were constructed at several of the major spring locations within NTS (Table 2) for the stage, freight, and mail lines that had begun to operate between Southern California and Utah. Hartwell and Valentine (2002, p. 61) identified three routes through current NTS lands – two along the ephemeral stream beds of *Beatty Wash* and *Fortymile Canyon,* and a third (the so-called *"Stagecoach Road"*) running east-to-west, connecting the southern foothills of Yucca Mountain and *Steves Pass* (in *Crater Flat*). All of these trail routes can be located on the 1986 metric edition of the U.S. Geological Survey (USGS) 1:100,000-scale topographic map for *Beatty, Nevada-California* quadrangle. Relay station construction frequently included improvements to the spring-discharge points. However, the significance of these sites was short-lived because of fluctuations in mining activity regionally and the establishment of other, more direct, thoroughfares to the West.

To encourage agricultural development of the West, the Federal government implemented a series of homesteading policies – e.g., the *Desert Lands Act of 1877*, the *Carey Act of 1894*, and the *Federal Reclamation Act of 1902* (Reisner, 1986) – intended to achieve reclamation of arid lands through irrigation-based farming. However, as noted in Volume 1 of this NUREG series, these

policies have had limited success in the Amargosa Desert area, for a variety of reasons. To the extent that any homesteading did take place in the area before 1900, it was confined mostly to discrete locations outside of current NTS boundaries, in the lower reaches of the valley. This literature review did not identify reports or other archeological evidence to suggest that homesteading (specifically farming) took place within the confines of current NTS boundaries.

Nevertheless, it is likely that some limited ranching may have taken place at some of these sites, including wild horse or mule herding, since corrals and barbed wire fences are reported in the literature. Long (1950, pp. 112–113, 116) provides a detailed account of activity in and around the *Emigrant Valley* and Amargosa Desert areas in the late 1930s-early 1940s, and noted that some ranching (cattle and sheep) was taking place at *White Rock Spring, Tippipah Spring, Cottontail Tank,* and the *Sheahan Ranch.* However, the ranchers and their families lived in near-by communities outside of present NTS boundaries (Fehner and Gosling, 2000; p. 10). Because water supplies were limited to cold springs, ranchers had to modify spring discharge points and build storage tanks to collect spring water (Worman, 1969). Also, as previously noted, Mendenhall (1909) recommended that travelers passing through the area carry adequate supplies of grain and hay for their horses because of inadequate forage at the lower elevations. It is likely that ranching took place at elevations above 4900 ft.[7]

---

[7] It should be noted that there is an extensive variety of vegetation throughout the site. However, the type of vegetation depends on elevation (temperature); slope; slope orientation; precipitation (climate); and soil properties. See Romney and others (1973); Beatley (1974, 1975); and Collins and others (1982). In general, steep slopes, especially those that face south or west at NTS, have little or no vegetation. Desert scrub (mesquite, salt grass, greasewood, and rabbit brush) which makes poor livestock forage, can be found at elevations of less than 4900 ft. At elevations of 4900 ft and above, higher-density cover and better livestock forage occur. At elevations of about 5900 ft and above, piñon pines, Joshua trees, and grasses begin to dominate.

Table 2. Possible Relay Stations within NTS. This table identifies the probable nearby sources of water supply.

Location	Latitude	Longitude	Water-Supply Source	Comments	Reference(s)
Captain Jack Spring	37° 10' 06" N.	116° 10' 12" W.	*Captain Jack Spring*	Small corral.	Worman (1969, p. 40)
Cane Spring	36° 47' 56" N.	116° 05' 45" W.	*Cane Spring*	1 stone and 2 frame cabins, corrals.	Long (1950, pp. 112–113); Worman (1969, pp. 12–15)
Fortymile Canyon	Multiple locations. See table below.			3 rock shelters.	Long (1950, p. 186); Anonymous (1969); Pippin and Zerga (1983, p. 54); Stoffle and others (1990, pp. 118–119)
Tippipah Spring	37° 02' 34" N.	116° 12' 13" W.	*Tippipah Spring*	2 stone cabins; stable, corrals, and barbed-wire pasture fence.	Long (1950, p. 112–113); Worman (1969, pp. 10–11)
Topopah Spring [a]	36° 59' 19" N.	116° 16' 17" W.	*Topopah Spring*	Ranch debris from fire (*ca.* 1951).	Anonymous (1969, pp. 7–8); Worman (1969, pp. 15–16); Brady (1975, p. 9)
White Rock Spring [a]	37° 12' 04" N.	116° 07' 04" W.	*White Rock Spring*	Cabin and corral.	Worman (1969, pp. 36–40)

a. Author inference based on literature.

Fortymile Canyon Springs [b]

Name	Location
unidentified	Section 16, T12S, R50E
Twin Springs (at Water Pipe Butte)	Section 27, T11S, R50E
Yellow Rock Spring	Section 23, T11S, R50E

b. *Black Spring* (?) and *Belted Mountain Spring* also shown on Ball's 1907 map.

There were some indigenous game animals in the area at the time so it is also likely that some of the abandoned sites may have also been used as lodges for hunting excursions.

2.3 Mining [8]

Some of the more common mineral resources identified in Southern Nevada include mercury, lead, zinc, and uranium. These resources occur frequently along faults and fracture zones in volcanic rocks. Despite extensive prospecting within NTS, no major occurrences of these metals have been reported. This conclusion would be consistent with regional mineral resource assessments of the area, including NTS (e.g., Schalla and Johnson, 1994; Sherlock and others, 1996). In general, most major mining activity in Southern Nevada has taken place outside of current NTS boundaries in the *Bare Mountain*, *Bullfrog-Rhyolite*, *Goldfield*, and *Johnnie Mining Districts*. See Lincoln (1923), Kral (1951), and Tingley and others (1993).

The first reported mining claim inside current NTS boundaries was in the *Timber Mountain Mining District* (Area 17), in 1869 (Angel, 1958; p. 486). Several prospects and mine shafts are depicted in the 1962 edition of the 1:24,000-scale USGS topographic map for the *Thirsty Canyon, Southeast* quadrangle, which includes this area. Although some chrysocolla, copper, and silver ores were produced between 1905 and 1917, this mining district's production was not considered significant. Other USGS topographic sheets covering current NTS boundaries depict additional small,

unrecorded prospect pits and claims, some of which have unspecified production. Bell and Larson (1982) describe some of these locations. In general, most of these locations were worked or staked before acquisition of the site by the Government and were not economically viable.[9] There are other examples of significant prospecting and mining that took place within NTS boundaries, as noted below:

- Ball (1907, pp. 128–130) reported prospects being developed in the *Oak Spring Mining District* (in Area 15) for precious metals and polymetallics as early as 1905; mining claims were reported as early as 1889 (McLane, 1996, p. 137). McLane also reports that the Bower Family took up residence near *Oak Spring* in 1920, building two stone ranch houses and establishing a mining company (*El Picacho*) to work near-by copper deposits between 1922 and 1928.[10] Scheelite ore containing tungsten ($CaWO_3$) and molybdenite ($PbMoWO_3$) was discovered in the *Climax Group* in 1936 or 1937, followed by development and production in the early 1940s (Kral, 1951); but significant tungsten production did not begin until the 1950s. After 10 years of co-use during the period of nuclear testing, the principal mining claims (*Climax* and *Crystal Mines*) were acquired through routine condemnation procedures and subsequently closed, because of the accidental releases of radiation to the site during the *Hard Hat* (1962), *Tiny Tot* (1965), and *Pile*

---

[8]Only *metallic resources* are discussed here. A large variety of *non-metallic resources* (industrial rocks and minerals) are present in large quantities throughout Southern Nevada. They include principally ceramic silica, zeolites, allunite, fluorite, sand, gravel, and lightweight construction aggregate (volcanic cinders, perlite, and pumice). See Nevada Bureau of Mines (1964). In general, these materials are found close to the ground surface and can be mined using near-surface, open-pit methods. Because the mining methods for these materials typically do not require water, the occurrence of non-metallic mineral resources is not addressed in this report.

---

[9]Approximately 123 mining claims are purportedly located on or near Yucca Mountain (Raney, 1988; p. 1) and only a few of these have been actually staked within the boundaries of the proposed repository site. In 1989, the staked claims were sold to DOE (Castor and others, 1990; p. 5).

[10]The Bower camp was ultimately abandoned about 1926 and was reported to be used in the early 1930s by criminal fugitives from Utah and Arizona (*Op cit.*).

Driver (1966) detonations (ERDA, 1977; p. 2-11). Overall, mining returns were small and frustrated by the incorporation of the claims into the Federal reservation.¹¹ The total extraction of ore was between 7 and 8 [short] tons containing 50 percent WO₃ concentrate (Johnson and Hibbard, 1957; pp. 380–381).

- A mercury mine and retort, at *Mine Mountain* (in Area 6), are frequently cited in the literature; claim notices indicate exploration in 1928 (Cornwall, 1972; p. 39). A number of prospects, shafts, and adits can be found in the 1961 edition of the 1:24,000-scale USGS topographic quadrangle map that bears the same name. Mercury was recovered from the mineral cinnabar occurring in seams and spheres in silicified and opalized rhyolite tuff (Cornwall and Kleinhampl, 1961), but no significant production is reported.

- In Area 26, high-grade silver-gold deposits were discovered in the *Wahmonie Mining District* in 1928 (Hewett and others, 1936; p. 71). The mining district attracted a population of between 1500 and 2000, supported by an extensive tent city comprised of hotels, saloons, a post office, and other commerce (Paher, 1980; pp. 322–324). Initial claims were mined-out in about 3 months. Despite the staking of about 1500 claims over an area of about 2 mi², and the development of multiple underground shafts (one at least 500 ft), no new ores were discovered (Miller, 1979; p. 157). As a consequence, the mining district was mostly abandoned by 1929.

- In the *Calico Hills* portion of Area 25, several prospect pits can be found on the 1961 edition of the 1:24,000-scale USGS topographic map for the *Jackass Flats* quadrangle. Castor and others (1990, p. 3) suggest that these particular workings are about the same age (*ca.* 1905) as workings at the *Horn Silver mine* (in the *Wahmonie Mining District*). Myrick (1992, p. 494) notes that the Calico Hills prospects were worked, without production, by the *Quartz Gold Mining Co.*, an affiliate of the *Sante Fe Railroad*.

- At the northwest end of Yucca Mountain, in Township 11 South, Range 48 East (T11S, R49E), slightly to the west of the NTS boundary, a small amount of mercury was produced in Section 29, at the *Thompson mine* (Bailey and Phoenix, 1944; p. 144). There was more moderate production of ceramic silica at the *Silica Mine* (*ca.* 1918–29), in Section 19 (Kral, 1951; p. 68). The 1954 edition of the 1:63,500-scale USGS topographic map for the *Bare Mountain* quadrangle shows these mines.

- Base metals were discovered in the *Groom Lake Mining District* about 1864, slightly to the east of the current NTS boundary (Humphrey, 1945). However, the mining district was not surveyed until about 1915. Most of the principal mining activity took place at the *Sheahan Groom Mine* from 1918–42, with limited mining until the early 1950s (Tschanz and Pampeyan, 1970; p. 148).

Table 3 summarizes the principal mining districts and identifies water supply sources. Overall, mineral exploration and mining had no impact on the development of water resources within NTS. The existing network of springs provided the mining districts with the supplies needed to support the scale of the respective operations.

15

Table 3. Reported Mining Districts within NTS.

Mining District	Latitude	Longitude	Water-Supply Source	Comments	Reference(s)
Groom Lake	37° 21 ' N.	115° 46 ' W.	Not reported. However, several springs reported in the mining district – *Disappointment*, *White Blotch*, *NaQuinta*.	Polymetallic replacement deposits mined periodically from 1864 to 1955; largest mine is the Groom mine (T7S, R5SE), acquired by the *Sheahan Family* in 1885.	Humphrey (1945, pp. 13, 35–45); Long (1950, p. 120); Tschanz and Pampeyan (1970, pp. 148–149)
Mine Mountain	37° 00 ' N.[a]	116° 07 ' W.[a]	Not reported.	Mercury retort associated with adits and shafts in T11S, R52E.	Cornwall (1972, p. 39); Worman (1969, p. 8); Brady (1975, p. 10).
Oak Spring	37° 14 ' N.	116° 13 ' W.	Initially *Oak Spring*, 900 ft southwest of the *Horseshoe claim*. Later, a 351-ft-deep well, 7 mi east of mine.	Climax mine worked tungsten skarn deposit until the early 1960s; other polymetallic claims reported to be worked in mining district before the Second World War.	Kral (1951, p. 138); Cornwall (1972, p. 39); McLane (1996, p. 138)
Wahmonic	36° 49 ' N.	116° 49 ' W.	*Cane Spring*	Comstock vein-type of deposits – *Horn Silver Mine*, *ca.* 1905 (in T5S, R47E).	Hewett and others (1936; p. 71); Brady (1975, pp. 8–9)

a. Author inference based on literature.

16

The exception was the *Oak Spring District*. Local springs were not sufficient for milling operations. Kral (1951, p. 138) noted that to provide the district with an adequate water, the *Goldfield Consolidated Mines Company* drilled four exploration wells locally. Water was located 7 mi to the east, at 291 ft below the ground surface. To get the water to the place of need, it was also necessary to lift the water an additional 860 ft over the *Rhyolite Hills* promontory.[11]

2.4   Land Withdrawals: 1930s–1950s

Most of the lands in Southern Nevada have historically been public. Establishment of a national system of parks in the United States resulted in the 1935 dedication of the *Desert National Wildlife Refuge*, in which the first 2482 mi^2 adjacent to current NTS boundaries were withdrawn from development and established for public use. However, it wasn't until the early 1940s that land within current NTS boundaries was withdrawn from public use. Originally, 640 mi^2 of public lands were withdrawn to create an aerial bombing and gunnery range for the Army Air Corps (formerly the *Las Vegas Bombing and Gunnery Range*; presently NAFR) in October 1940 (Fehner and Gosling, 2000; p. 20). At the time, more than 90 percent of the area was in the public domain. Brady (1975, p. 7) reports that between 300 and 400 cattle openly ranged at *Topopah*, *White Rock*, and *Cane Springs* before the war, when native grasses were taller and more plentiful. This type of activity would be consistent with available historic and archeological information. However, there were a number of grazing, mining, and homesteading claims on the land which made taking government possession difficult. In August 1941, the Government began condemnation proceedings against the outstanding parcels of land. In October of the same year, the courts finalized the land condemnations and

federal marshals cleared any remaining stragglers off of the range (Fehner and Gosling, 2000; p. 20).

2.5   Atomic Programs

At the end of the war, training activities at NTS were briefly discontinued. Some cattle ranching is reported to have returned, but only at *Topopah Spring*, *Frenchman Flats*, and *Emigrant Valley* (Anonymous, 1969; pp. 7–8). One individual obtained a grazing lease to two-thirds of the test area and ran about 40 horses and 250 cattle (Fehner and Gosling, 2000; p. 51). Table 4 lists the grazing locations, along with the water supply sources. However, at the time, there was growing concern about the strategic vulnerability of the United States atomic testing programs in the Pacific Ocean (*Op cit.*, pp. 39–40). Relying on a top-secret feasibility study, code-named *Nutmeg* and conducted by the Pentagon, the AEC selected the Las Vegas Bombing and Gunnery Range as the site of the United States continental nuclear test site in December 1950.[12] Control of the site was later assumed by the AEC. Open-range cattle ranching was allowed well into the 1950s, in the Yucca Flat area (*Op cit.*, p. 51) provided certain provisions were met.[13]

---

[11]Using Kral's description, scaling the reported 7-mi distance to a 1:250,000-scale geologic map of Southern Nye County (Cornwall, 1972) would place the 350 ft, 8-in cased well in alluvial deposits of Emigrant Valley.

[12]There were several reasons for the selection. Foremost among these were the vast open spaces, within closed topographic basins at the site, needed to minimize the risks of atmospheric testing, as well as favorable climatic conditions (ERDA, 1977; pp. 2-12 – 2-13). Other important criteria included the remoteness of NTS and the fact that the site was already under Federal control.

[13]Although developed to allow mining within the *Oak Spring Mining District*, the following provisions were also applied to ranching operations on-site (Boyer, 1952): (a) that all ranching personnel have the appropriate security clearance; (b) all ranching personnel would evacuate affected areas during atomic tests; (c) access to NTS from the north would discontinue; (d) the AEC retained the right of unfettered site inspections; (e) ranching personnel would have to adhere to the same regulations as NTS workers; and (f) ranch owners could not file claims for damages resulting from [these] restrictions or government operations.

Table 4.   Possible Ranching Locations within NTS.

Location	Latitude	Longitude	Water Supply Source	Comments	Reference(s)
Bowen Ranch	37° 14' 23" N.	116° 02' 30" W.	Local springs (*Oak* and *Tub Springs*) provided sufficient water for domestic use.	Two stone cabins and corral. Site later served as base for mining operations.	McLane (1996)
Emigrant Valley			*Cottontail Tank*	Open grazing in Southern Belted Mountains area.	Long (1950, pp. 116–117)
George Lathrop Ranch		*Lathrop Well* (in vicinity of Highway 95 and State Road 373 intersection)[a]	Concrete water tank at *Lathrop Well*; water hauled by wagon from *Fairbanks Spring* (36°29' N, 115°20' W); water tank used in earlier construction of *Las Vegas and Tonopah Railroad.*[b]	Before the Second World War, livestock grazed in Jackass Flats (NTS Area 25).	Fehner and Gosling (2000, p. 10)
Sheahan (or Sheehan) Ranch	37° 26' N.[c]	116° 53' W.[c]	Exact supply of water not reported; *Cattle Springs* closest water supply (?).	Emigrant Valley ranch operated until 1955. Now referred to as "Area 51." Elevation ≈ 6400 ft.	Long (1950); Worman (1969, p. 8); Solnit (1994)
Tippipah Spring	37° 02' 34" N.	116° 12' 13" W.	*Tippipah Spring*	2 stone cabins, stable, corrals, and barbed-wire pasture fence. Occupied late 1930s-early 1940s.	Long (1950, pp. 112–113); Worman (1969, pp. 10–11)
Topopah Spring[d]	36° 59' 19" N.	116° 16' 17" W.	*Topopah Spring*	Ranch debris from fire (*ca.* 1951) – *NaQuinta Ranch* (?).	Anonymous (1969, pp. 7–8); Worman (1969, pp. 15–16); Brady (1975, p. 9)
White Rock Spring[d]	37° 12' 04" N.	116° 07' 04" W.	*White Rock Spring*	Cabin and corral. Occupied during late 1930s-early 1940s.	Worman (1969, pp. 36–40)

a. Kensler (1982) reports (unspecified) historic ruins at Lathrop Wells.
b. Ken Garey, *Bar-B-Q Ranch* (Amargosa Valley), personal communication, November 1998.
c. Approximate location.
d. Author inference based on literature.

Ultimately, water and grazing rights for the remaining ranches and leases were acquired by negotiated purchase in 1955 (ERDA, 1977; p. 2-12). Many of the cattle purchased subsequently became part of a government-owned herd that was allowed to range freely until the early 1960s.[14] Further land withdrawals by the AEC to the west (in 1954) and the north (in 1964) established the current dimensions of NTS (Figure 1). Today, under the direction of DOE, NTS is approximately 1350 mi^2 and is surrounded by a 4120 square mile-buffer zone (that includes NAFR and the *Tonopah Test Range*).

Over the years there have been a number of atomic programs and activities within NTS, and they have had the greatest influence on the development of water supply within the site. The first atmospheric nuclear test at the NTS, named *ABLE*, took place at Frenchman Flat on January 27, 1951. After the *Limited Test Ban Treaty of 1963*, nuclear tests were conducted underground, based in large measure on the conclusion that there were favorable (and extensive) subsurface geologic conditions at the site (Houser, 1968; p. 21). In addition to the approximately 900 nuclear weapons tests, other notable atomic programs and activities at NTS have included nuclear rocket experiments, the *Plowshare Program* and the *MX (Peacekeeper)* mobile missile siting studies. Since the nuclear weapons testing moratorium in 1992, test site use has diversified into many other programs such as hazardous chemical spill testing, emergency response training, conventional weapons testing, and waste management and environmental technology studies.

In the late 1950s and early 1960s, the geohydrology of the area was subject to detailed study for the first time to better understand the potential effects of atomic testing activities at NTS on regional ground-water supplies. These studies were conducted by the USGS on behalf of the then

AEC. In addition to this work, the USGS supported the Nevada Department of Conservation and Natural Resources in the conduct of basin-wide, ground-water resource analyses for the state. After the creation of the United States' HLW management program, in the 1980s, there was a renewed interest in the ground-water hydrology of NTS because of the focus on Yucca Mountain as a candidate site for a potential geologic repository. Since then, there have been hundreds of new geohydrologic studies added to the literature.[15]

To support its mission, an infrastructure has been created within NTS over time to provide the necessary services, including construction of the Mercury base camp in the fall of 1950. Originally, Mercury was a temporary staging site used to support atomic testing in nearby Frenchman Flat. It was initially designed to be used no more than 18 weeks per year, and included buildings to house scientific, communications and computing equipment, weather monitoring installations, and radiological safety facilities. Residential barracks were present as well as a mess hall and an administrative center. At the time, Mercury had a peak housing capacity of about 412 personnel (Fehner and Gosling, 2000; p. 81). As it became apparent that testing would become more routine at the site, the camp's capacity was expanded to 1100, and included the construction of additional barracks, a second mess hall, a recreation facility, warehouses, and laboratories (*Op cit.*, pp. 81–82).

This infrastructure has been described by DOE (1996; pp. 4-19 — 4-25), including information on water supply (*Op cit.*, pp. 4-22, 4-24). Most of the major geologic investigations of NTS included the sinking of one or more exploratory and test wells,

---

[14]In 1963, through an interagency agreement, the cattle came under the jurisdiction of EPA's radionuclide environmental monitoring program. See Appendix A.

[15]Slate (1999) includes an annotated bibliography compiled by Morrow and Machette (1999) that provides a comprehensive summary of the literature on the local and regional hydrogeology of Southern Nevada, including NTS.

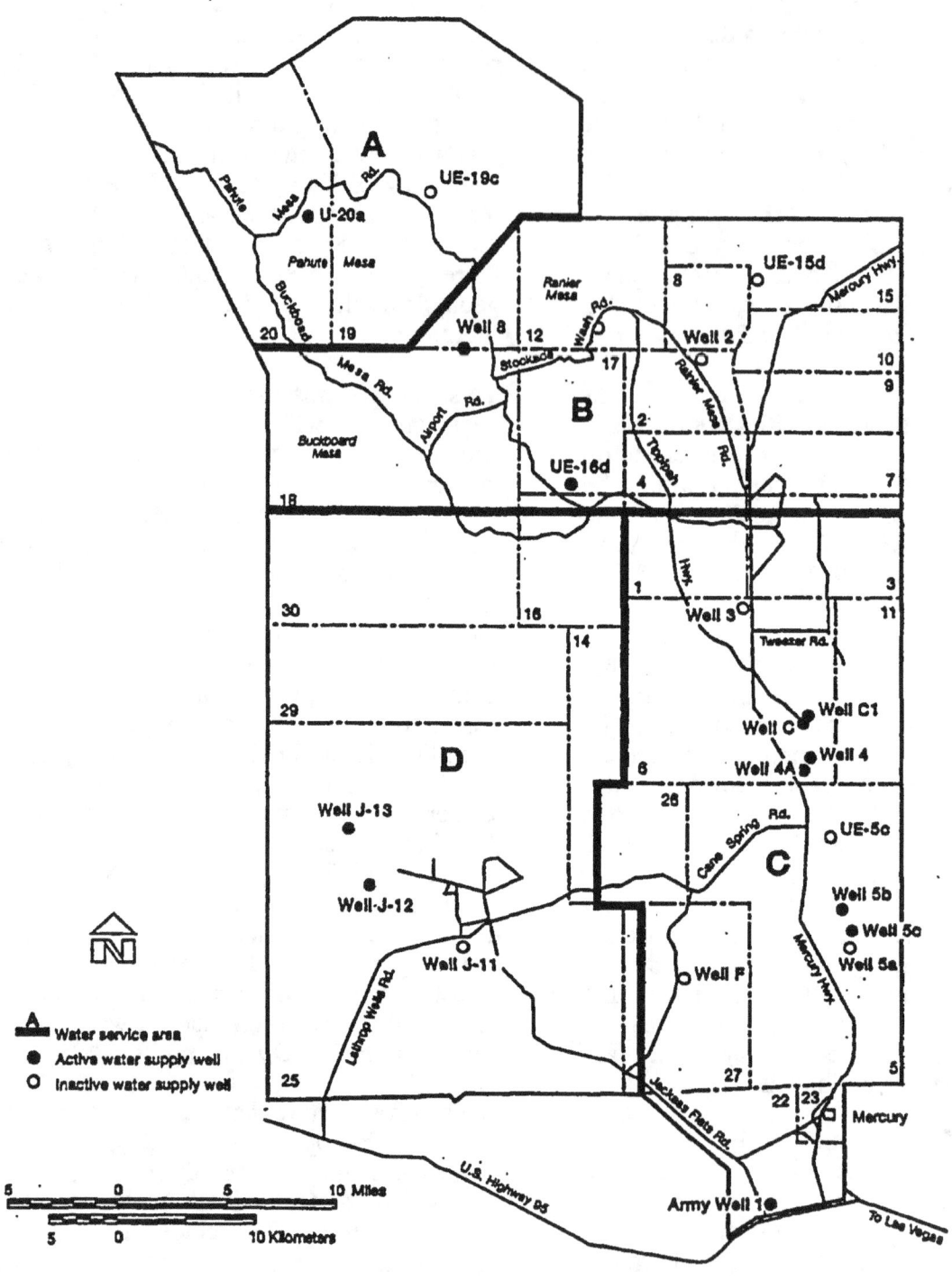

Figure 5. **Wells within NTS.** Activities and water supply wells within these areas are described in the appendices. Taken from DOE (1996, p. 4-25).

sometimes leading to the development of a production well. As a result, today seventeen wells supply the freshwater needs for NTS (Figure 5). Most of the wells are located in *Yucca Flat*, *Frenchman Flat*, and *Mercury Valley*, and were drilled in the mid-to-late 1950s or early 1960s, using conventional cable tool and rotary-air drilling technology. Wells were drilled by private contractors, with USGS oversight. NTS water-supply needs are presently met by 12 of these wells — 11 of the 12 wells are potable. Cold springs within NTS are not sufficient to meet the water supply needs. The first water-supply well supporting this infrastructure was *Army Well No. 1*. It is located in Mercury Valley, near Highway 95 (Section 4 of T16S, R53E). The 1253-ft well was completed in May 1956.[16] This well initially met the fresh water needs for *Camp Desert Rock* [17] and later, the Mercury base camp, which it still supports today.

In contrast to wells in the Amargosa Desert area to the south, which pump ground water from a shallow alluvial aquifer (depths of 100 ft or less), most NTS ground water is pumped from relatively deep volcanic or regional carbonate aquifers (depths of 1000 ft or more). Also, unlike alluvial aquifers, where ground water is contained in a porous granular matrix, carbonate and volcanic rocks are essentially non-granular media so the amount of ground water available from these aquifers depends on the presence of hydraulically conductive faults, fractures, and joint sets (e.g., California Department of Water Resources, 1991). These networks of structural features within the rock mass

control the movement of ground water.

Claassen (1973), citing other references, contains descriptions of many of the NTS well designs. All the wells are lined, and perforated well casings act as screens. Casing diameters decrease telescopically with depth, and range from 24 in (outside diameter), at the ground surface, to 4¾ in, at depth. None of the wells is operating at full capacity, and only one, Army Well No. 1, has a capture zone that extends beyond current NTS boundaries. Because the combined capacity of these wells exceeds demand, many wells are used on a rotating basis. NTS has six water transmission systems, totaling about 100 mi, that distribute the water supply among four service areas (Appendix B). Potable water is piped underground, whereas non-potable water is piped above ground. Pipe diameters range from 4 to 12 in. To meet peak demand, this distribution system is complemented by 27 storage tanks. In 1996, five additional water storage tanks were under construction (DOE, 1996; p. 4-20). Potable water (including bottled water) is trucked to support facilities and operations that are not connected to this system.

Peak historic water demands within NTS are shown in Table 5. In 1994, NTS used 450 million gallons of water (DOE, 1996; p. 4-23). The Mercury site was the primary consumer, accounting for about 40 percent of total use (*Op cit.*). In general, the effects of well pumping at NTS have resulted in the lowering of water levels in the vicinity of the particular water supply well and some localized changes in ground-water flow directions. Pumping effects are mostly limited to within a few thousand meters of the operating wells (*Op cit.*, p. 4-117). As shown by Table 5, the peak water demand associated with historic NTS activities has been a small fraction of the estimated perennial yield of most of the affected individual hydrographic areas.

For the purposes of DOE's Yucca Mountain site characterization programs, Wells J-12

---

[16]See the NDWR URL site at *www.ndwr.state.nv.us/IS/wlog/wlog.htm*.

[17]Camp Desert Rock was a military installation established in NTS Area 22 to house the U.S. Army's Atomic Maneuver Battalion participating in military operations during atomic testing. It was under the command of the U.S. Sixth Army, Presidio, San Francisco, California. At the peak of its operations, Camp Desert Rock had 100 semi-permanent buildings, more than 500 tents, a 7500 ft runway, and some 6000 troops in residence.

Table 5. Perennial Yields and Peak Historic Water Demand for NTS Hydrographic Areas. Taken from DOE (1996, p. 4-118). Hydrographic areas are shown in Figure 2.

NDWR Hydrographic Area	Estimated Perennial Yield (10^6 gallons/yr)	DOE Water Supply Wells [a]	Peak Historic Demand	
			10^6 gallons	Year
Gold Flat	6100	1	1100	1989
Kawich Valley	7100	1	1400	1989
Emigrant Valley	8200	None [b]	No DOE Demand	
Yucca Flat	1100	8	2600	1989
Frenchman Flat	50,000	3	1200	1962
Mercury Valley	25,900	1	?	1992
Rock Valley	25,900	None [b]	No DOE Demand	
Fortymile Canyon	25,000	3	1100	1988
Oasis Valley	6600	None [b]	No DOE Demand	
Amargosa Valley	76,600	None [b]	No DOE Demand	

a. Additional information on these wells can be found in the appendices.
b. No DOE water supply wells are located in this basin.

and J-13, located in Jackass Flats, are the current sources of water supply. These two wells (and a third well, J-11) were drilled in the early 1960s, when Jackass Flats was the site of the Nuclear Rocket Development Station (NRDS – see Young, 1972).[18] Fresh water was needed for NRDS facility construction, test-cell and exhaust cooling, and thermal shielding. To meet NRDS program needs, the time, a welded tuff aquifer – the Topopah Spring Member of the Paintbrush Tuff. The average consumption of water for NRDS in 1967 was 522,000 gpd, and peak consumption was 720,000 gpd – all of which was supplied by Well J-13 (Young, 1972; p. 18). Well J-12, as the stand-by well, had an additional capacity of 1 million gpd (*Op cit.*). Well J-11 was ultimately abandoned because of the poor chemical quality of the water (sulfate content of about 480 mg/L) and the fact that casing corrosion had reduced well yield. Completion details for Wells J-12 and J-13 can be found in Figures 6 and 7, respectively.

[18]Located roughly in the center of Area 25, NRDS occupied about 123 mi^2 and was selected as the site for a series of ground tests of reactors, engines, and rocket stages for possible use in the Nation's space program. (This program was first initiated in the early 1950s at this site under the name *Project Rover*.) NRDS was jointly administered by AEC and the National Aeronautics and Space Administration's Nuclear Propulsion Office. NRDS consisted of three test cells, an engine test stand, and reactor and engine maintenance assembly and disassembly facilities, as well as other administrative, support, and storage areas. The program ended in 1972. During its operational life, more than $100 million was spent on NRDS facility construction and equipment, including the development of water supplies (Anonymous, 1993a; pp. 22–23).

An additional well that could be used as a potential source of ground-water supply is Test Well JF-3, also in Jackass Flats, located about 2600 ft south of Well J-12 (Plume and La Camera, 1996). It was completed in the Topopah Spring tuff in 1992. The well penetrated 479 ft of alluvium and 817 ft of volcanic rock. The average hydraulic conductivity of the tuffs beneath the water table was estimated to exceed 330 ft/day.

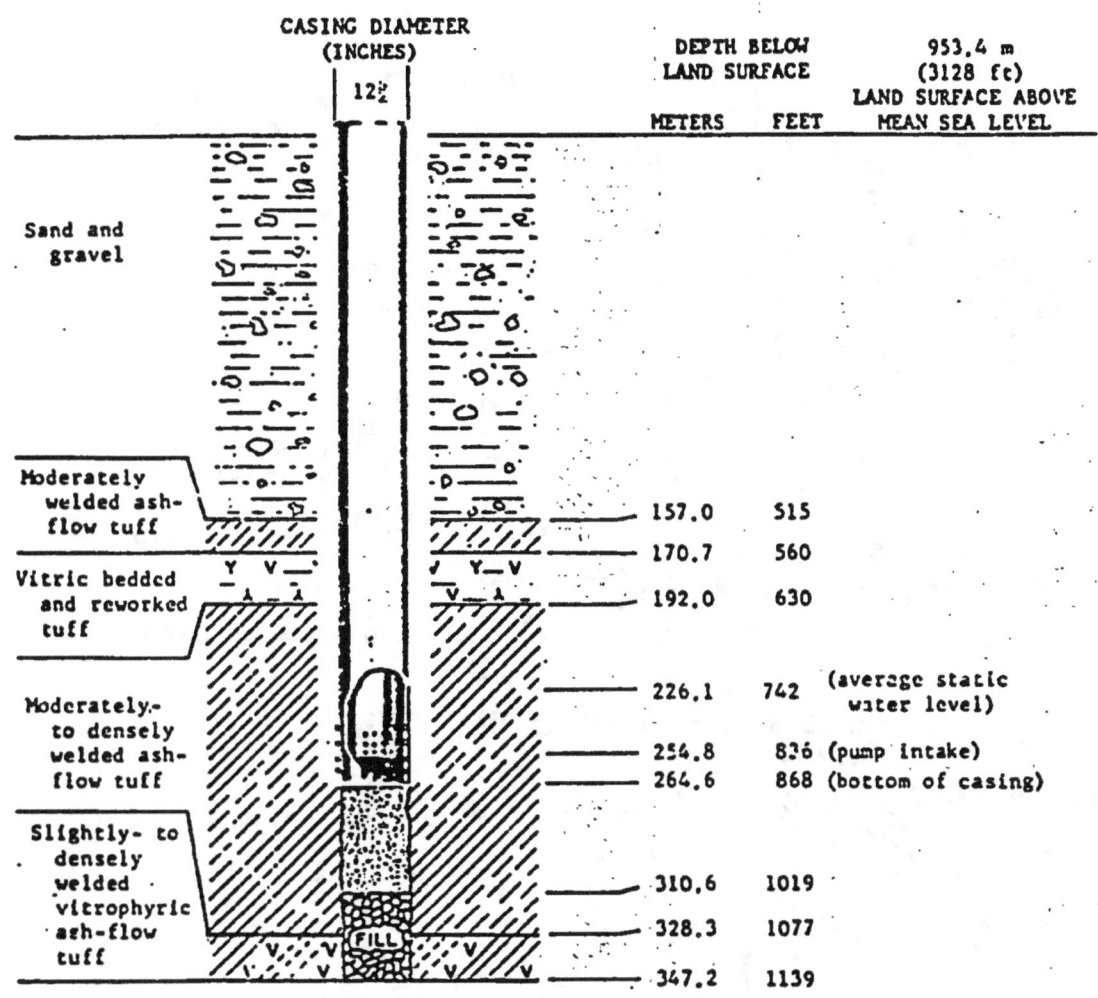

Figure 6.   Completion Details for Well J-12. Includes lithology. Taken from Claassen (1973, p. 24).

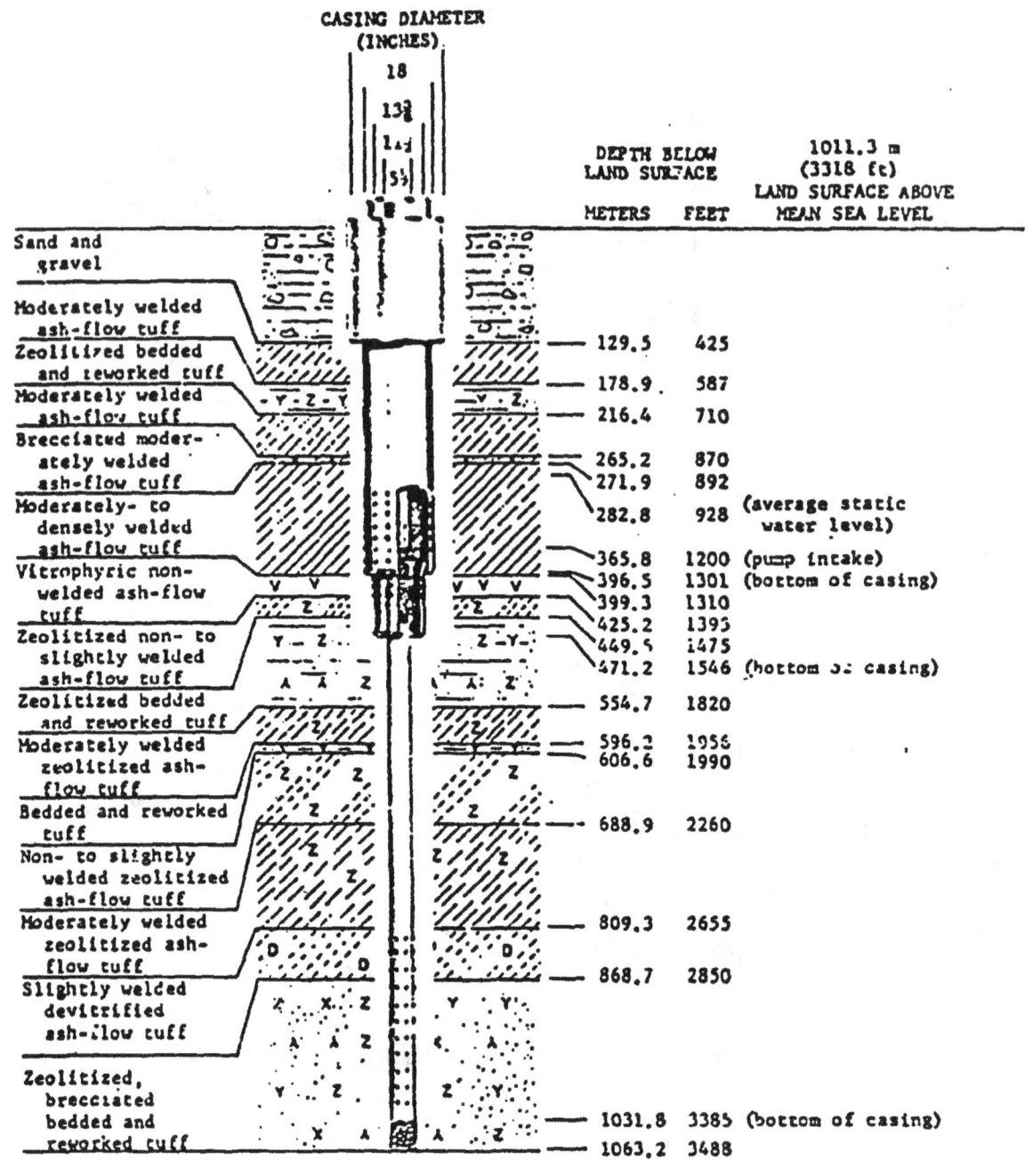

Figure 7.   Completion Details for Well J-13. Includes lithology. Taken from Claassen (1973, p. 32).

24

3. REFERENCES

Angel, M., *Thompson and West's History of Nevada, 1881*, Berkeley, Howell-North, 1958.

Anonymous, "NTS History Steeped in Fact and Fiction," *NTS News*, 12(16):4-8 [August 8, 1969].

Anonymous, "Uniqu e 36-Acre Experimental Farm Tested Crops, Animals," *NTS News and Views* [April 1993]. [Special edition.]

Baker, T.L., S.R. Rae, J.E. Minor, and S.V. Connor, "Water for the Southwest – Historical Survey and Guide to Historic Sites," Fairfax, American Society of Civil Engineers (ASCE), *ASCE Historical Publication No. 3*, 1973.

Bailey, E.H., and D.A. Phoenix, "Quicksilver Deposits in Nevada," Reno, University of Nevada Bulletin, Vol. 38, No. 5, *Geology and Mining Series*, No. 41, December 1944.

Ball, S.H., "A Geologic Reconnaissance in Southwestern Nevada and Eastern California," U.S. Geological Survey Bulletin 308, 1907.

Beatley, J.C., "Effects of Rainfall and Temperature on the Distribution and Behavior of *Larrea tridentata* (Creosote-bush) in the Mojave Desert of Nevada," *Ecology*, 55(2):245-261 [1974].

Beatley, J.C., "Climates and Vegetation Patterns Across the Mojave/Great Basin Desert Transition of Southern Nevada," *The American Midland Naturalist*, 93(1):53-70 [1975].

Bell, E.J., and L.T. Larson, "Overvi ew of Energy and Mineral Resources for the Nevada Nuclear Waste Storage Investigations, Nevada Test Site, Nye County, Nevada," Las Vegas, Reynolds Electrical and Engineering Co., Inc., NVO-250, September 1982. [Prepared for DOE's Nevada Operations Office.]

Boyer, M.W., General Manger, Nevada Test Site, Atomic Energy Commission, Memorandum to Lt. General L.A. Pick, Department of the Army, Washington, D.C. [General Subject: "Acquisition of Privately Owned Land and Mining Rights"], July 18, 1952.

Brady, W.J., "The Early Days at NTS," *NTS News/ Bicentennial Issue 1975*, 5(5):7-11 [1975].

Bryan, K., "Classification of Springs," *Journal of Geology*, 27:522-561 [1919].

California Department of Water Resources, "Ground Water in Fractured Hard Rock," Sacramento, *Water Resources Facts No. 1*, April 1991.

Castor, S.B., S.C. Feldman, and J.V. Tingley, "Mineral Evaluation of the Yucca Mountain Addition, Nye County, Nevada," Reno, Nevada Bureau of Mines and Geology [NBMG], NBMG Open File Report 90-4, 1990.

Claassen, H.C., "W ater Quality and Physical Characteristics of Nevada Test Site Water Supply Wells," Lakewood, U.S. Geological Survey, USGS-474-158 [NTS-242], 1973. [Non-serial report prepared for the AEC.]

Collins, E., T.P. O'Farrell, and W.A. Rhoads, "Biologic Overview for the Nevada Nuclear Waste Storage Investigations, Nevada Test Site, Nye County, Nevada," Goleta, EG&G Energy Measurements, EGG 1183-2460, January 1982. [Prepared for the Sandia National Laboratories.]

Cornwall, H.R., and F.J. Kleinhampl, "Geology of the Bare Mountain Quadrangle, Nevada," U.S. Geological Survey Quadrangle Map GQ-157, 1 sheet, scale 1:62,500, 1961.

Cornwall, H.R., "Geology and Mineral Deposits of Southern Nye County, Nevada," Reno, University of Nevada/Mackay School of Mines, Nevada Bureau of Mines and Geology Bulletin 77, 1972.

Dohrenwend, J.C., "Basin and Range," in W.L. Graf (ed.), "Geomorphic Systems of North America," Boulder, *Geological Society of America Centennial Special Volume 2*, 1987.

Drollinger, H., C.M. Beck, R. Furlow, and others, "Cultural Resources Management Plan for the Nevada Test Site," Las Vegas, Desert Research Institute, DOE/NV11508-47, June 1999. [Prepared for DOE's Nevada Operations Office.]

Energy Research & Development Administration, "Fin al Environmental Impact Statement: Nevada Test Site, Nye County, Nevada," Washington, D.C., ERDA-1551, September 1977.

Fehner, T.R., and F.G. Gosling, "Origins of the Nevada Test Site," U.S. Department of Energy, DOE/MA-0518, December 2000.

French, R.H., A. Elzeftawy, J. Bird, and B. Elliot, "Hydrology and Water Resources Overview for the Nevada Nuclear Waste Storage Investigations, Nevada Test Site, Nye County, Nevada," Las Vegas, Desert Research Institute, NVO-284, June 1981. [Issued June 1984.]

Geraghty, J.J., D.W. Miller, F. van der Leeden, and F.L. Troise, *Water Atlas of the United States,* Port Washington, Water Information Center Publication, 1973.

Hartwell, W.T., and D. Valentine, "Reading the Stones: The Archaeology of Yucca Mountain," Las Vegas, Desert Research Institute, Topics in Yucca Mountain Archaeology No. 4, 2002. [Prepared for DOE.]

Hevesi, J.A., A.L. Flint, and L.E. Flint, "Simulation of Net Infiltration and Potential Recharge Using a Distributed-Parameter Watershed Model of the Death Valley Region, Nevada and California," U.S. Geological Survey Water-Resources Investigations Report 03-4090, 2003.

Hewett, D.F., E. Callaghan, B.N. Moore, T.B. Nolan, W.W. Rubey, and W.T. Schaller, "Mineral Resources Around the Region of Boulder Dam," U.S. Geological Survey Bulletin 871, 1936.

Houser, F.N., "Application of Geology to Underground Nuclear Testing, Nevada Test Site," in E.B. Eckel (ed.), "Nevada Test Site," Boulder, *Geological Society of America Memoir 110,* 1968.

Humphrey, F.L., "Geology of the Groom District, Lincoln County, Nevada," Reno, University of Nevada Bulletin, Vol. 39, No. 5, *Geology and Mining Series,* No. 42, June 1945.

Johnson, M.S., and D.E. Hibbard, "Geology of the Atomic Energy Commission Nevada Proving Grounds, Nevada," U.S. Geological Survey Bulletin 1021-K, 1957.

Kensler, C.D., "Survey of Historic Structures: Southern Nevada and Death Valley," San Francisco, URS/John A. Blume & Associates, Report JAB-00099-121, July 1982. [Prepared for DOE's Nevada Operations Office.]

Kral, V.E., "Mineral Resources of Nye County," Reno, University of Nevada Bulletin, Vol. 45, No. 3, *Geology and Mining Series, No. 50*, January 1951.

Lincoln, F.C., *Mining Districts and Mineral Resources of Nevada*, Reno, Nevada Newsletter Publishing Co., 1923. [1982 reprint by Nevada Publications.]

Lee, M.P., N.M. Coleman, and T.J. Nicholson, "History of Water Development in the Amargosa Desert Area, Nevada: A Literature Review," U.S. Nuclear Regulatory Commission, NUREG-1710, Vol. 1, February 2005.

Lingenfelter, R.E., *Death Valley & the Amargosa: A Land of Illusion*, Berkeley, University of California Press, 1986.

Long, M., *The Shadow of the Arrows – Death Valley 1849–1949*, Caldwell, Caxton Printers, 1950. [Revision to the 1941 edition].

Manly, W.L., *Death Valley in '49*, Chicago, Lakeside Press, 1927.

McCracken, R.D., *A History of Amargosa Valley, Nevada*, Tonopah, Nye County Press, 1990.

McCracken, R.D., *The Modern Pioneers of the Amargosa Valley*, Nye County Press, 1992.

McLane, A.R., "El Picacho —The Writing Cabin of B.M. Bower," *Nevada Historical Society Quarterly*, 39(2):134–146 [Summer 1996].

Mendenhall, W.C., "Some Desert Watering Places in Southeastern California and Southwestern Nevada," U.S. Geological Survey Water-Supply Paper 224, 1909.

Miller, R.E., "Welcoming Address for Rocky Mountain Section, Geological Society of America," in E.B. Eckel (ed.), "Nevada Test Site," Boulder, *Geological Society of America Memoir 110*, 1968.

Miller, D.C., *Ghost Towns of Nevada*, Boulder, Pruett Publishing, 1979.

Morrow, P.J., and M.N. Machette (compilers), "Selected References for the Geology of Death Valley National Park and Surrounding Region," Denver, U.S. Geological Survey, 1999. [Compilation prepared for the "Conference on Status of Geologic Research and Mapping in Death Valley National Park," held in Las Vegas, Nevada, on April 9-11, 1999.

Myrick, D.L., *The Railroads of Nevada and Eastern California: Part II – The Southern Railroads*, Reno, University of Nevada Press, 1992.

Nevada Bureau of Mines, "Mineral and Water Resources of Nevada," Reno, University of Nevada/ Mackay School of Mines, Bulletin 65, 1964. [Prepared by the USGS and the Nevada Bureau of Mines.]

Paher, S.W., *Nevada Ghost Towns and Mining Camps*, San Diego, Howell-North Books, 1980.

Pippin, L.C. and D.L. Zerga, "Cultural Resources Overview for the Nevada Nuclear Waste Storage Investigations, Nevada Test Site, Nye County, Nevada," Las Vegas, University of Nevada System, Desert Research Institute, NVO-266, November 1983. [Prepared for DOE.]

Plume, R. W. and R. J. La Camera, "Hydrogeology of Rocks Penetrated by Test Well JF-3, Jackass Flats, Nye County, Nevada," Carson City, U.S. Geological Survey Water-Resources Investigations Report 95-4245, 1996.

Raney, R.G., " Status of Lode-Quartz Claims on Yucca Mountain," Spokane, U.S. Bureau of Mines, September 1988. [Prepared for the NRC.]

Reisner, M., *Cadillac Desert - The American West and Its Disappearing Water,* New York, Viking-Penguin Books, 1986. [Revised 1993.]

Romney, E.M., V.Q. Hale, A. Wallace, O.R. Lunt, J.D. Childress, H. Kaaz, G.V. Alexander, J.E. Kinnear, and T.L. Ackerman, "Some Characteristics of Soil and Perennial Vegetation in Northern Mojave Desert Areas of the Nevada Test Site," Los Angeles, University of California, USAEC Report UCLA12-916, June 1973. [Prepared for the AEC.]

Rush, F.E., "Index of Hydrographic Basins," Carson City, Nevada Department of Conservation and Natural Resources, Water Resources Information Series Report 6, September 1968.

Schalla, R.A., and E.H. Johnson (eds.), *Oil Fields of the Great Basin*, Reno, Geological Society of Nevada, 1994.

Sherlock, M.G., D.P. Cox, and D.F. Huber, "Known Mineral Deposits and Occurrences in Nevada (Chapter 2)," in D.A. Singer (ed.), "A n Analysis of Nevada's Metal-Bearing Mineral Resources," Reno, Nevada Bureau of Mines and Geology, Open File Report 96-2, 1996. [Prepared in cooperation with the USGS.]

Slate, J.L. (ed.), "Proceedings of Conference on the Status of Geologic Research and Mapping, Death Valley National Park," U.S. Geological Survey Open-File Report 99-133, 1999. [Includes a diskette of references compiled by Morrow and Machette.]

Solnit, R., *Savage Dreams: A Journey into the Hidden Wars of the West*, San Francisco, Sierra Club Books, 1994.

Steward, J.H., " Basin-Plateau Aboriginal Sociopolitical Groups," *Smithsonian Institution Bureau of American Ethnology Bulletin*, Vol. 120, 1938. [1997 reprint by the University of Utah Press.]

Stoffle, R.W., D.B. Halmo, J.E. Olmsted, and M.J. Evans, "Native American Cultural Resource Studies at Yucca Mountain, Nevada," Ann Arbor, Institute for Social Research, The University of Michigan, 1990. [Originally prepared for DOE by Science Applications International Corporation.]

Thordarson, W., and B.P. Robinson, "Wells and Springs in California and Nevada within 100 Miles of the Point 37 Deg. 15 Min. N., 116 Deg. 25 Min. W., on Nevada Test Site," U.S. Geological Survey, USGS-474-85, 1971. [Prepared for the AEC.]

Tingley, J.V., R.C. Horton, and F.C. Lincoln, "Outlin e of Nevada Mining History," Reno, University of Nevada/Mackay School of Mines, Nevada Bureau of Mines Special Publication 15, 1993.

Tschanz, C.M., and E.H. Pampeyan, "Geology and Mineral Deposits of Lincoln County, Nevada," Reno, University of Nevada/Mackay School of Mines, Nevada Bureau of Mines and Geology Bulletin 72, 1970.

TRW Environmental Safety Systems, Inc., "Summary of Socioeconomic Data Analyses Conducted in Support of Radiological Monitoring Program during Calendar Year 1994," Las Vegas, Document No. DE-AC01-91RW001134, June 1995. [Prepared for DOE.]

U.S. Department of Energy, "Draft Environmental Assessment: Yucca Mountain Site, Nevada Research and Development Area, Nevada," Office of Civilian Radioactive Waste Management, DOE/RW-0012, December 1984.

U.S. Department of Energy, "Environmental Assessment: Yucca Mountain Site, Nevada Research and Development Area, Nevada," Office of Civilian Radioactive Waste Management, 3 vols., DOE/RW-0073, May 1986.

U.S. Department of Energy, "Site Characterization Plan, Yucca Mountain Site, Nevada Research and Development Area, Nevada," Office of Civilian Radioactive Waste Management, DOE/RW-0199, 9 vols., December 1988.

U.S. Department of Energy, "Final Environmental Impact Statement for the Nevada Test Site and Off-Site Locations in the State of Nevada," Nevada Operations Office, DOE/EIS-0243, 3 vols. (and appendices), August 1996.

U.S. Department of Energy, "Final Environmental Impact Statement for a Geologic Repository for the Disposal of Spent Nuclear Fuel and High-Level Radioactive Waste at Yucca Mountain, Nye County, Nevada," Office of Civilian Radioactive Waste Management, DOE/EIS-0250, 4vols., February 2002.

U.S. Environmental Protection Agency, "Environmental Radiation Protection Standards for Yucca Mountain, Nevada — Draft Background Information Document for Proposed 40 CFR 197," Office of Radiation and Indoor Air, EPA 402-R-99-

008, August 1999.

Waddell, R.K., Jr., J.H. Robison, and R.K. Blankennagel, "Hydrology of Yucca Mountain and Vicinity, Nevada-California — Investigative Results through Mid-1983," U.S. Geological Survey Water-Resources Investigations Report 84-4267, 1984.

Walker, G.E., and T.E. Eakin, "Geology and Ground-Water of Amargosa Desert, Nevada – California," Carson City, Nevada Department of Conservation and Natural Resources, Ground-Water Resources Reconnaissance Series Report 14, 1963.

Worman, F.C.V., "Archeological Investigations at the U.S. Atomic Energy Commission' Nevada Test Site and Nuclear Rocket Development Station," Los Alamos, Los Alamos Scientific Laboratory, LA-4125, August 1969. [Prepared for the AEC.]

Young, R.A., "Water-supply for Nuclear Rocket Development Station at the Atomic Energy Commission's Nevada Test Site," U.S. Geological Survey Water-Supply Paper 1938, 1972.

APPENDIX A

MAJOR NEVADA TEST SITE (NTS) FACILITIES AND WATER SUPPLY SOURCES

NTS AREA	PRINCIPAL ACTIVITY
Nuclear Test Zone	100 atmospheric detonations before August 1993 *Limited Test Ban Treaty*; about 800 underground nuclear tests before 1992 nuclear weapons testing moratorium.
High-Explosive Test Zone	61 underground nuclear tests between 1957–92.
Reserved Area	Controlled-access area that provides a buffer between defense and non-defense research, development, and testing activities; Department of Defense currently operates a high explosives research & development facility.
Critical Assembly Zone	Area used for nuclear explosive operations — staging, assembling, modification.

NUCLEAR TEST ZONE

Location	Water Supply Source	Comments	Reference(s)
Yucca Flat weapons test basin	See Appendix B.		DOE (1996, pp. 4-10 – 4-17)
Area 1	27 mi²	Nuclear testing from 1952–90.	
Area 2	20 mi²	Nuclear testing from 1952–90.	
Area 3	32 mi²	Nuclear testing from 1952–92.	
Area 4	16 mi²	Nuclear testing from 1952–91.	
Area 7	20 mi²	Nuclear testing from 1964–91.	
Area 8	13 mi²	Nuclear testing from 1958–88.	
Area 9	20 mi²	Nuclear testing from 1951–92.	
Area 10	20 mi²	Nuclear testing from 1961–91 under *Plowshare Program.*	
Frenchman Flat test basin			
Area 6	82 mi²	Nuclear testing from 1957–90.	
Plutonium Valley test basin			
Area 11	26 mi²	Includes some "Reserved Area"	
Pahute Mesa			
Area 19	150 mi²	Nuclear testing from mid-1960s – 1992.	
Area 20	100 mi²	Nuclear testing from mid-1960s – 1992.	

Location		Water Supply Source	Comments	Reference(s)
HIGH-EXPLOSIVE TEST ZONE				
Rainier Mesa		See Appendix B		DOE (1996, pp. 4-10 – 4-17)
Area 12	40 mi²			
RESERVED AREA				
Area 5	95 mi²	See Appendix B.		DOE (1996, pp. 4-10 – 4-17)
Area 11	26 mi²			
Area 14	26 mi²			
Area 15 *U.S. Environmental Protection Agency Bioenvironmental Experimental Farm*	37 mi²	Rehabilitated well UE-15d (and 3 acre-ft reservoir).	3 underground tests conducted during 1962-65.	DOE (1996, pp. 4-10 – 4-17)
			27-acre farm and dairy operated from 1964–81. 30 Holstein cows, 100 Hereford beef cattle, and other horses, pigs, goats, and chickens raised on farm-grown forage and vegetables. Site included 15 acres of agricultural, plots, 2 acres of microplots, and irrigated greenhouse. Elevation ≈4501 ft.	Anonymous (1993, p. 26); ERDA (1977, pp. 2-17, 2-137); DOE (1996, p. 4-15)
Area 17 *Animal Investigation Program*	31 mi²	Water pond at Test Well 8 (Timber Mountain Moat area ª).	Small herds of beef cattle (75 to 100 head) used for on-site radiological surveillance from 1955 through the 1970s. Elevation 5696 ft.	ERDA (1977, pp. 2-135 – 2-137)
Area 18	90 mi²	See Appendix B.	Site of 5 nuclear weapons tests (1962– 65) and Plowshare Program cratering tests.	DOE (1996, pp. 4-10 – 4-17)
Area 22 *Camp Desert Rock*	32 mi²	Pre-existing *Las Vegas* and *Tonopah Railroad* wells.	ca. 1951–57.	Anonymous (1993, pp. 2, 4); DOE (1996, pp. 4-15 – 4-16)
Area 23 *Mercury*	5 mi²	Wells 5B and 5C (Frenchman Flats) and Army Well 1 (Mercury Valley).	1951– present.	DOE (1996, p. 4-16)
Area 25	223 mi²			DOE (1996, pp. 4-10 – 4-17)
Nuclear Rocket Development Station (NRDS)		Wells J-11, J-12, and J-13 (Jackass Flats).	Mid-1950s to 1972.	Young (1972)

Location		Water Supply Source	Comments	Reference(s)
Rock Valley Study Area		None.	Controlled-study area selected in 1960 for desert ecosystem studies.	DOE (1996, p. 4-17)
Yucca Mountain Site Characterization Program		Wells J-12 and J-13 (Jackass Flats).	Early 1980s-present; potential geologic repository for high-level radioactive waste.	Young (1972); DOE (1988)
Area 26	22 mi²	See Appendix B.	*Project Pluto* nuclear ramjet tests.	DOE (1996, p. 4-17)
Area 29	62 mi²	See Appendix B.		
Area 30	58 mi²	See Appendix B.	Limited nuclear testing; site of 1968 *Operation Buggy* as part of Plowshare Program.	DOE (1996, p. 4-17)

CRITICAL ASSEMBLY ZONE

Location		Water Supply Source	Comments	Reference(s)
Area 27	50 mi²	See Appendix B.		DOE (1996, p. 4-17)

a. And possibly other areas in which there had been nuclear testing activities.

References

Anonymous, "Unique 36-Acre Experimental Farm Tested Crops, Animals," *NTS News and Views* [April 1993]. [Special *NTS News and Views* edition.]

Energy Research & Development Administration [ERDA], "Final Environmental Impact Statement: Nevada Test Site, Nye County, Nevada," Washington, D.C., ERDA-1551, September 1977.

U.S. Department of Energy, "Chapter 3. Hydrology," in "Site Characterization Plan, Yucca Mountain Site, Nevada Research and Development Area, Nevada," Office of Civilian Radioactive Waste Management, DOE/RW-0199, Vol. II, Part A, December 1988.

U.S. Department of Energy, "Final Environmental Impact Statement for the Nevada Test Site and Off-Site Locations in the State of Nevada," Las Vegas, Nevada Operations Office, DOE/EIS 0243, 3 vols. (and appendices), August 1996.

Young, R.A., "Water-supply for Nuclear Rocket Development Station at the Atomic Energy Commission's Nevada Test Site," U.S. Geological Survey Water-Supply Paper 1938, 1972.

APPENDIX B

SUMMARY OF WELL INFORMATION FOR NEVADA TEST SITE SERVICE AREAS

Adopted from DOE (1996, pp. 4-21 – 4-22, 4-133), unless otherwise noted.

Well	Location	Service Area(s)	Aquifer Type	Water Quality	Well Depth (ft)	Static Water Level (ft)	Pump Depth Setting (ft)	Yield [a] (gpm)	Status
				SERVICE AREA A					
UE-19e	Area 19	19, 20 (?)	Volcanic	Not reported	6005 [b]	2218 [b]	2300 [b]		Inactive
UE-19c	Area 19	19, 20	Volcanic	Nonpotable	8489			359	Inactive
UE-19 g-s	Area 19	19, 20 (?)	Volcanic	Not reported	7506 [b]	2044 [b]	2183 [b]		Inactive
UE-20a [d]	Area 20	19, 20	Volcanic	Nonpotable	4500 [b]	2066 [b]	2302 [b]	280	Active
				SERVICE AREA B					
Well 2	Area 2	2, 4, 7, 9, 10	Carbonate	Potable (with chlorination)	3423 [b]	2055 [b]	2814 [b]	169	Inactive
Well 8	Areas 18 and 19	2, 12	Volcanic	Potable	5490	1073	1228	399	Active
UE-1r	Area 1	1		Nonpotable					Inactive
UE-15d [b]	Area 15	2, 4, 7, 8, 9, 10, 12, 15, 17, 18	Volcanic	Potable (with chlorination) [d]	5940	668	1659	270	Inactive
UE-16D	Area 17	1	Carbonate	Potable	3000	755	1083	193	Active

B-1

Well	Location	Service Area(s)	Aquifer Type	Water Quality	Well Depth (ft)	Static Water Level (ft)	Pump Depth Setting (ft)	Yield[a] (gpm)	Status
SERVICE AREA C									
Army Well No. 1	Area 22	22, 23	Carbonate	Potable (with chlorination)[d]	1946	690	951	531	Active
Well 3	Area 6	3	Alluvial	Nonpotable	1799[b]	1595[b]	1717[b]	none	Inactive
Well 4	Area 6	6	Alluvial	Potable	1479	941	1271	650	Active
Well 4a	Area 6	6	Volcanic	Potable				230	Active
Well 5A	Area 5	27	Volcanic	Nonpotable	910[b]	707[b]	781[b]	90	Abandoned
Well 5B	Area 5	5, 22, 23	Alluvial	Potable	900	684	753[b]	270	Active
Well 5C	Area 5	5, 22, 23	Alluvial	Potable	1187	693	693	325	Active
Well A	Area 3	3 (?)	Alluvial	Potable (with chlorination)[d]	1870[b]	1614[b]	1823[b]	161[c]	Inactive
Well C	Area 5	3, 6	Carbonate	Potable	1701	1544	1553	270	Active
Well C1	Area 5	3, 6	Carbonate	Potable	1707	1548	1591	280	Active
Well F	Area 27	27		Nonpotable				238	Inactive
UE-5c[e]	Area 5	5	Alluvial	Nonpotable				349	Inactive
SERVICE AREA D									
J-11[e]	Area 25	25	Volcanic	Nonpotable	1432	1120		none	Inactive
J-12	Area 25	25	Volcanic	Potable	1139	798[d]	822	817	Active
J-13	Area 25	25	Volcanic	Potable	3755[d]	999[d]	1151	679	Active

a. Well yields calculated from controlled pump tests are typically within one order of magnitude of well drillers' estimates.
b. Claassen (1973).
c. DOE (1988, p. 3-124).
d. Construction water and environmental sampling.
e. Young (1972).

References

Claassen, H.C., "Water Quality and Physical Characteristics of Nevada Test Site Water Supply Wells," Lakewood, Colorado, U.S. Geological Survey, USGS-474-158 [NTS-242], 1973.

U.S. Department of Energy, "Chapter 3. Hydrology," in "Site Characterization Plan, Yucca Mountain Site, Nevada Research and Development Area, Nevada," Office of Civilian Radioactive Waste Management, DOE/RW-0199, Vol. II, Part A, December 1988.

U.S. Department of Energy, "Final Environmental Impact Statement for the Nevada Test Site and Off-Site Locations in the State of Nevada," Nevada Operations Office, DOE/EIS 0243, 3 vols. (and appendices), August 1996.

Young, R.A., "Water-supply for Nuclear Rocket Development Station at the Atomic Energy Commission's Nevada Test Site," U.S. Geological Survey Water-Supply Paper 1938, 1972.

RC FORM 335
-2004)
RCMD 3.7

U.S. NUCLEAR REGULATORY COMMISSION

BIBLIOGRAPHIC DATA SHEET

(See Instructions on the reverse)

1. REPORT NUMBER
(Assigned by NRC, Add Vol., Supp., Rev., and Addendum Numbers, if any.)

NUREG-1710, Volume 2

. TITLE AND SUBTITLE

History of Water Development at the Nevada Test Site: A Literature Review

3. DATE REPORT PUBLISHED

MONTH	YEAR
February	2005

4. FIN OR GRANT NUMBER

. AUTHOR(S)

M.P. Lee and N.M. Coleman

6. TYPE OF REPORT

Technical

7. PERIOD COVERED *(Inclusive Dates)*

. PERFORMING ORGANIZATION - NAME AND ADDRESS *(If NRC, provide Division, Office or Region, U.S. Nuclear Regulatory Commission, and mailing address; if contractor, provide name and mailing address.)*

Advisory Committee on Nuclear Waste Staff
U.S. Nuclear Regulatory Commission, Washington, D.C. 20555-0001

. SPONSORING ORGANIZATION - NAME AND ADDRESS *(If NRC, type "Same as above"; if contractor, provide NRC Division, Office or Region, U.S. Nuclear Regulatory Commission, and mailing address.)*

Same as above.

0. SUPPLEMENTARY NOTES

1. ABSTRACT *(200 words or less)*

Historic accounts, geologic treatises, and other key literature sources were used to chronicle developments in the Nevada Test Site (NTS) during the past 150 years. As was the case in the nearby Amargosa Desert, human activities in the area currently occupied by NTS were initially influenced by the location of cold springs. They provided indigenous Native Americans with drinking water. Later, as part of the Western expansion, many of these same springs were relied on by Euro-American pioneers as they crossed the continent. By the time NTS was engaged in activities related to the Nation's defense, it was necessary to develop the available subsurface ground-water supplies, aided in part by improved geologic knowledge of local resources. The first well supporting this infrastructure was Army Well No. 1. The 1253-foot well was completed in May 1956. Today, 17 wells distributed among four service areas supply NTS water needs. Most were drilled in the mid-to-late 1950s or early 1960s in Yucca Flat, Frenchman Flat, and Mercury Valley. Overall, the welded volcanic tuff aquifer is only locally important (in Jackass Flats) whereas the lower carbonate aquifer serves other portions of NTS.

This report is the second volume in the NUREG-1710 series.

12. KEY WORDS/DESCRIPTORS *(List words or phrases that will assist researchers in locating the report.)*

ground water
mining
Native Americans
Nevada Test Site
springs
water resources
well drilling
Yucca Mountain

13. AVAILABILITY STATEMENT

unlimited

14. SECURITY CLASSIFICATION

(This Page)

unclassified

(This Report)

unclassified

15. NUMBER OF PAGES

16. PRICE

NRC FORM 335 (9-2004)

PRINTED ON RECYCLED PAPER

Printed
on recycled
paper

Federal Recycling Program

FEBRUARY 2005

HISTORY OF WATER DEVELOPMENT AT THE NEVADA TEST SITE: A
LITERATURE REVIEW

NUREG-1710, Vol. 2

UNITED STATES
NUCLEAR REGULATORY COMMISSION
WASHINGTON, DC 20555-0001

OFFICIAL BUSINESS

www.ingramcontent.com/pod-product-compliance
Lightning Source LLC
Chambersburg PA
CBHW081908170526

45167CB00007B/3203